# VULTURE'S
# ROW

# VULTURE'S ROW

## THIRTY YEARS IN NAVAL AVIATION

### Paul T. Gillcrist

**Schiffer Military/Aviation History**
Atglen, PA

## ACKNOWLEDGEMENTS

The characters in these stories are those aviators with whom I flew as a naval aviator for thirty-three years. They ran the full gamut of personalities . . . from extremely colorful to extremely bland. Sure, there were some with warts. There were one or two whom I did not respect. However, there are some generalities I can make about most of them, which my counterparts in the private sector, unfortunately, can never make.

They were all dedicated to flying. They were all, in their own way, gutsy. They all told the truth in matters professional; and therefore could be trusted in the air. They all were possessed with a sense of personal accountability for their own actions. A few of them put my life in jeopardy, but never with malice aforethought. Many more of them saved my life . . . and when push came to shove, could be relied upon to go to great extremes to put themselves on the line for me.

But, many of the characters in these stories were not aviators. Some were plane captains, or catapult crewmen, or flight deck directors, or maintenance personnel. For these people . . . they would say they were just doing their jobs! But, in doing those jobs, no sacrifice was too great.

These characters made up the greatest fraternity in the world. I am proud to be a part of that fraternity.

---

*Dedication*
*To my son, Jim . . . who does everything so well. You make me proud!*

---

*Front dust jacket: Modern day Landing Signal Officer's platform "waving the Tomcat" aboard the USS NIMITZ, WESTPAC 1987.*

*Rear dust jacket: "Paddles" waving a Cougar aboard USS ORISKANY (CVA-34), Sea of Japan, 1954.*

*Page 1: VF-192 F9F-5 Panther on the port catapult of the USS ORISKANY, Sea of Japan, 1955.*

*Page 2-3: F-14A at the ramp aboard USS NIMITZ, 1987.*

*End sheets: VF-193 F2H-3N Banshee "in the wires" aboard USS ORISKANY, Sea of Japan, 1955.*

Book Design by Robert Biondi.

Copyright © 1996 by Paul T. Gillcrist.
Library of Congress Catalog Number: 95-72353.

Printed in China.
ISBN: 0-7643-0047-4

We are interested in hearing from authors with book ideas on related topics.

Published by Schiffer Publishing Ltd.
77 Lower Valley Road
Atglen, PA 19310
Please write for a free catalog.
This book may be purchased from the publisher.
Please include $2.95 postage.
Try your bookstore first.

# FOREWORD

Someone once described an aviation career as hours and hours of sheer boredom punctuated by brief moments of stark terror. Although the author has never found flying on and off of an aircraft carrier boring, I can attest to the occasional moments of excitement, exhilaration and, yes, stark terror. Since 27 years separated my first and last carrier landings, those accumulated "moments" represent a substantial bank of interesting events which I personally experienced or to which I was a witness.

This volume is a collection of some of the most interesting of those events. They have stayed fresh in my memory for a variety of reasons: some because they were funny, others because they were genuinely exciting, still others because they were beautiful or sad. But, all of them were carefully recorded in my memory with some details contained in my pilot's flight log book. Each flight is recorded by an entry in that book telling the date, the bureau number and type of airplane, and numerical details about flight time, all on a single line with a box at the right margin labelled simply, "Remarks." It is not a large box, but it is big enough for a few words. Of the several thousand entries in my log book there are four where a single explanatory word was sufficient to call to mind infinite detail about the flight they describe. In two of those four entries I had scribbled simply, "ejected." That was enough! In the other two the word I scribbled was "crashed." That was also enough!

I struggled with the idea of using the phrase "sea stories" in the subtitle. The reason being that to a navy man, a sea story is a description of an event which has been so embellished in the re-telling of it, that over the years, it bears little resemblance to the original version. In this volume, that is not the case. The stories are all original versions, and are as factual as I could record them when I wrote them down so many years ago in diary entries and letters to my family.

This volume is divided into eight parts. Each part represents a distinct and discrete element of my flying career. The stories contained in this volume are, in my mind, logically divided into these eight groupings, and I have attempted to provide the reader with an explanation, at the beginning of each part as to why I did it this way.

Every carrier pilot has a logbook full of memories. It is his flight logbook, the official record of his flying career, related in the form of rows and rows of numbers and letters. I am sure that each of us has pulled it out of the old cruise box where it gathers dust, and paged through it. The perusal never fails to evoke memories, powerful ones that will never fade. In this volume, I am giving the reader a brief look at some of the pages in my logbook.

A word about the title. Back in the early carrier days . . . before the pilot landing assessment television (PLAT) was available in all the ready rooms and elsewhere on the ship, the only place that anyone could watch carrier operations was "Vulture's Row." This was generally an area on the port side of the island structure several levels above the flight deck and near the after portion of the island. Depending on the class of carrier, it could be a gun tub, an external passageway or even a catwalk. The only requirement was an unobstructed view of the flight deck . . . particularly the landing area. This was where pilots who weren't flying could go with a camera and have a reasonable expectation of coming away from any launch or recovery with some exciting photos.

Every carrier pilot I know from that era has a dozen or so boxes of 35mm color slides he took from "Vulture's Row." That was where you could really capture the excitement of carrier aviation!

# CONTENTS

*Flight of USAF F-86s and VF-191 F9F-6s from Johnson AFB, Japan on April 29, 1955. (Official U.S. Navy photograph)*

PART I

# "SATAN'S KITTENS"

*"Give me operations,*
*Way out on some lonely atoll*
*For I am too young to die.*
*I just want to grow old."*

*Refrain from Army Air Corps drinking song from World War II.*

*The venue for this part of the volume is U.S. Navy Fighter Squadron VF-191, better known as "Satan's Kittens," one of the most famous squadrons in the entire history of U.S. naval carrier aviation. Satan's Kittens were the day fighter squadron in Carrier Air group 19 and we flew the Grumman F9F-6 Cougar. The night fighter squadron, VF-193, flew the McDonnell F2H-3N Banshees. The third jet squadron, VF-192, flew the Grumman F9F-5 Panthers and served in both a fighter and attack role. The only other squadron in the air group was the attack squadron VA-195, and they flew the A1-H Skyraiders.*

*Of course, there were detachments of night attack A-1s, heavy attack A2Js which also provided an aerial refueling capability to the night fighter squadron, and a photo detachment of F2H-1P Banshees. It was a very capable air group for that time frame, 1954.*

*When I began my career in carrier aviation in 1953, the United States Navy was in the throes of a major technological change . . . the switch from reciprocating to jet engine propulsion. This change had already begun in the field of both military and commercial aviation . . . but had not yet made any profound inroads into naval aviation.*

*U.S. carriers, which had survived the post-World War II disarming, were the same ones which had been the main battery of the fleet during the war in the Pacific. They were known as the "fast carriers." But, in those days they were what we know now as "straight-deck carriers." Without the dramatic post-war revolutionary changes of the angled flight deck, the optical landing system and the steam catapult, they were extremely limited in the context of jet propulsion. Those three changes, inherited from our*

British brothers, were the very upgrades which would permit the change to jet propulsion.

I am sure that history will show that the real heroes of carrier aviation were those intrepid young men who finally figured out how to operate jet-powered aircraft from carriers.

But the lesson was expensive. The attrition rate of carrier-based aircraft in 1953 was 25 times higher than it is today! In recent years, when the Soviet Navy tried to emulate our standard carrier operations from decks like the TIBLISI, they learned a hard lesson. We had been making it look easy when, in fact, routine, safe carrier operations are only achieved by constant practice and absolute adherence to a set of operating practices developed over the last 40 years.

Illustrative of this point is the ramp strike by Lieutenant Frank Repp during ORISKANY's 1954 WestPac deployment. Frank was a pilot in the air group's night fighter squadron, VF-193. On a reasonably clear day, Frank flew his F2H-3N Banshee into the ramp (we called it the "spudlocker"). The airplane broke apart on impact with the unyielding ship's structure. The only part of the airplane which came aboard was the first 10 feet of the cockpit area of the fuselage. Lieutenant (junior grade) Bill MacColl of VF-191, happened to be up in Vulture's Row at the time armed with his brand new Nikon camera and, by cycling the new high-speed shutter/rewind mechanism as fast as he could, he was able to capture the event in a series of ten spectacular color photographs.

The photo sequence displayed on the following pages shows the burgeoning ball of bright red fire and black smoke which followed Frank as he slid up the flight deck in the most brilliant color demonstration I have ever seen of the hazards of carrier aviation!

Gradually, Frank's fiery bit of flotsam slowed to a majestic halt as the ship's smoke eaters swarmed over it in a desperate attempt to save the pilot from immolation. Frank somewhat casually unstrapped himself and stepped out onto the flight deck, an unexplicable smile on his face.

Frank was one of the lucky pioneers of jet-powered carrier aviation!

*Pages 12-16: Lt. Frank Repp's ramp strike aboard USS ORISKANY (CVA-34) on 2 March 1954 in the Sea of Japan. (Photos by W. MacColl)*

# 1

## "RIP"

Lieutenant "Rip" Rohrer arrived as a new member of Fighter Squadron VF-191 a few months after I did. He was a second tour pilot, which meant that he had already had a tour of duty in a fleet squadron before reporting in as a new member of Satan's Kittens. His previous tour, however, had been in an East Coast fighter squadron and as a propeller or "prop" pilot flying F4U Corsairs.

His total jet experience was about 50 hours of jet transition training in the Lockheed two-seater, the T-33 Shooting Star. So, although he was not a "nugget" like me, his lack of jet experience and unfamiliarity with West Coast operations meant he would be treated like a nugget at least for a while.

Rip was a handsome, blond haired, blue eyed, lightly complexioned young man in his late 30s. He stood about five feet eight inches tall and was powerfully built. He had enlisted in the Navy in the early 1950s, and went into flight training from the enlisted ranks. During his time as an enlisted man, he had become the all-Navy middleweight wrestling champion.

Rip and I ended up on the flight schedule together on a familiarization/orientation flight for him. Because of his previous fleet experience, he was designated the section leader and I his wingman. We had just completed some basic aerobatics and two-plane aerial combat maneuvering in a training area northeast of Oakland, California, and were headed back toward Moffett Field. We were in a standard combat spread formation at 35,000 feet when I spotted two U.S. Air Force F-86 fighters approaching us from six o'clock at a distance of about five miles. They were obviously making a run on us. I called them out to Rip. He acknowledged seeing them and told me we would execute a Half Cuban Eight maneuver and meet them head on. A Half Cuban Eight was essentially an Immelman Turn (half a loop with a roll-out on top) and would have been an acceptable tactical maneuver had we been starting at no more than 15,000 feet. But there was absolutely no way an F9F-6 Cougar could do it starting at 35,000 feet . . . no way! I knew this basic aeronautical fact, but obviously, Rip did not. Before I could think of a diplomatic way of telling him so, Rip started up. In utter dismay I followed him through the start of this embarrassing tactical evolution.

After all, he was the leader; and a good wingman sticks with his leader . . . no matter what. . . and I was a good wingman! We made it just a little past the vertical when we ran out of airspeed and ideas at about the same time. The last time I looked at my airspeed indicator it read 90 knots. After that I didn't have the heart to look at it. I had even lost sight of our attackers and, as a matter of fact, had lost interest in them for the moment.

In his defense, I feel compelled to observe that Rip had the native intelligence to neutralize his flight controls and let the airplane fall though his "hammerhead" maneuver. I am certain our Air Force compatriots were a good 15,000 feet below us watching this remarkable display of airmanship with great amusement.

But I was in trouble! Although we had begun the maneuver from a combat spread formation (half a mile abeam of each other), somehow we had begun to drift toward one another, with no flight controls to alter our ballistic trajectory. By now we were at about 45,000 feet and falling out of the sky at zero airspeed. I could see that we were going to collide in another few seconds unless I did something. As the wingman, it was my responsibility to remain clear of my leader. The problem was that I had no control over my destiny . . . or at least I didn't think so. All I could think of was how humiliating it would be to pull a caper like this in full view of two other airmen.

In desperation I pushed the stick full forward and held it there. There must have been just a tiny bit of pitch authority left, because I felt the nose move up just a hair. My airplane passed within a few feet of the underside of his, just as it entered an inverted stall. Then, all of a sudden my airplane entered a flat, inverted spin.

"God damn it!," I recall shouting into my mask as I initiated inverted spin recovery procedures and held them. The airplane's attitude was essentially flat, the yaw rate was not too great, and I was hanging in the straps with all sorts of junk; dirt, bolts, pieces of safety wire and such coming out from underneath the ejection seat. Some of the crap got into my eyes.

Nothing else happened! I held inverted spin controls for a full five turns . . . to no avail. "Jesus," I thought to myself, "this is beginning to be not so funny." Next, I tried upright spin controls . . . and nothing happened. The airplane, I noticed, was falling past 25,000 feet when Rip called me. He told me what my altitude was and asked me if I was all right. My response was that I was okay, but my voice must

*Author in VF-191 Grumman F9F-6 Cougar, Air Group 19, USS Oriskany (CVA-34), 1955. (Author's collection)*

have contained a mixture of disgust and outrage. Next I tried inverted spin recovery techniques at full power . . . then at idle power. Then I extended the speed brakes. Through all of this I could feel the rudder pedals shaking. At one point I reached down and turned off the yaw damper, thinking that it might be holding me in the spin. Passing 15,000 feet I heard Rip call me.

"Two, you'd better think about getting out of that thing." I could think of nothing by way of rejoinder so I kept silent. All of a sudden, the airplane pitched abruptly into an upright spin, made two more turns and recovered in a steep nose-down attitude. I recall glancing at the altimeter after leveling out directly over the top of Mount Tamalpais. The instrument read about 6,000 feet. There was not much clearance between me and the top of that mountain.

The return flight to Moffett Field was relatively uneventful. I found myself thinking that although the Cougar is not supposed to remain in a steady-state inverted spin, my airplane had fallen out of control for 39,000 feet! It was the longest prolonged, out-of-control maneuver of my flying career up until then; and I didn't care to exceed that record ever again!

Rip was very apologetic during the post-flight de-briefing. I didn't say much because I blamed myself for being so stupid as to follow him through a maneuver that was, by the laws of physics, unexecutable. But I also made one of those vows,

*"Satan's Kittens" (VF-191) F9F-6 Cougars preparing for launch aboard the USS Oriskany, Sea of Japan, 1955. (Official U.S. Navy photograph)*

which if kept, keep aviators alive over the years. I promised myself that I would never, ever let a section leader put me in extremis again! It was a valuable lesson . . . one that I have never forgotten!

Rip was still apologizing when we left the squadron ready room at quitting time. He offered to buy me a conciliatory drink. Since it was Friday afternoon, and "happy hour" time, I accepted. There was simply no way to stay angry at "Rip." He had such a pleasant manner and a winning grin that he could charm the socks off anybody . . . even our stern Executive Officer, "Tiny" Graning. Activity was picking up at the Moffett Field Officer's Club when we arrived.

My squadron running mate, Buck Baillie, and I had a double date later that evening with a couple of nurses. Rip's family happened to be out of town and he was at loose ends so I foolishly invited him to join us later on. Naturally, he accepted.

When Buck and I arrived at the Officer's Club an hour or so later with our dates, Rip joined us. It became immediately apparent that Rip had enjoyed more than a few drinks since I left him at Happy Hour. We all went into the dining room and sat down for dinner. Rip ordered the biggest steak on the menu but didn't participate in the lively conversation going on at the table while we sipped our drinks and waited for dinner to arrive.

Right in the middle of an animated discussion I was having with my date, I was interrupted by a loud gargling noise and glanced over at Rip sitting on my right. To my utter astonishment, I saw that Rip had fallen asleep face down in his plate. The strange noise was the sound of his breathing through the gravy.

This put a damper on the conversation and I realized that we couldn't ignore him. Buck helped me get Rip over my shoulder in a fireman's carry and I made my way out of the dining room with as much aplomb as I could muster up under the circumstances. What I didn't know was that as I left the table with Rip on my shoulder, he regained enough of his senses to retrieve the steak off his plate. With each step I took, the gravy soaked piece of meat, dangling from Rip's outstretched hand flapped against the back of my sports jacket.

Getting Rip's 155 pounds in and out of my car, up a flight of stairs to my BOQ room and into the spare bed was not easy. Nonetheless, fifteen minutes later I was back at our table in a cleaner sports jacket and a more pleasant mood. The evening took a turn for the better after that.

An unusual sound woke me the next morning. I rolled over in my bed and saw Rip sitting cross-legged in his bed, clad only in boxer shorts, tearing with his teeth at the steak he held in his hands. "Good morning," he said with a broad grin. Then, holding the steak out toward me, he asked, "Do you want a bite?" I almost threw up!

The following Monday morning the entire squadron deployed to Naval Auxiliary Air Station, Fallon, Nevada, for a three-week gunnery detachment. It was great fun! We lived in our flight suits, flew gunnery missions twice a day, slept in World War II open-bay barracks, played the slot machines and drank too much at the casinos in the small town of Fallon, just sixty miles east of Reno.

It was the following Friday at Happy Hour at the air station's small officer's club when Rip added yet another page to the growing body of folklore surrounding this colorful character. He was sitting at a booth in the club's bar, taking on all comers in arm wrestling. Everyone in the room was at least vaguely aware that his latest challenger, the air wing Operations Officer, Lieutenant Commander G.G. Smith was having a go at it. "G.G.," as he was called, was a rather diminutive, slightly built fellow who looked as though he had never done a push-up in his life. What possessed him to take on the U.S. Navy's middleweight wrestling champion in arm wrestling escaped me. Perhaps it was the Jack Daniels he had been drinking that gave him the courage.

At any rate, the level of noise in the bar was high. Voices raised in loud conversations, the clicking of billiard balls, the wailing of western music on the jukebox; all were suddenly pierced by a sharp, snapping sound, not unlike the sound made by the snapping of a large twig. All eyes turned to the booth where Rip and G.G. were sitting. The latter's face was as white as a sheet; as he stared unbelievingly at his arm; his hand still enveloped by Rip's huge fist. G.G.'s lower arm was lying flat on the table while the upper arm was twisted 90 degrees past its normal limits. The upper arm bone, the fibula, had suffered an extreme torsion fracture. G.G. was taken to the Naval Hospital in Oakland for extensive restorative bone surgery.

The remainder of the gunnery deployment was relatively uneventful. Rip had elected to drive his own automobile from Moffett because he needed it to haul all of his extensive prospecting equipment. He had acquired a Geiger counter and a sizable amount of other prospecting equipment; and had spent most of his free time roaming the rugged mountains surrounding Fallon in search of a uranium lode. He took off from Fallon on the last day of the deployment with a car-load of rock samples, headed south for Moffett Field.

One of his squadron mates, following along several hours behind him, came across Rip sitting by the side of the road through Tioga Pass on top of a wooden box of rock samples looking a little disconsolate. Alongside him was an enormous pile of prospecting gear and his luggage. The good Samaritan stopped and was helping Rip load all of his equipment into his car, when he asked, "What happened, Rip?"

"The son of a bitch quit on me for the last time," came the prompt reply.

"You mean your old jeep?" the Samaritan asked.

"Yep," was the laconic response.

"Where is it?" the Samaritan asked looking around.

"Down there," Rip answered with a nod of his head.

The squadron mate, who was loading the last of Rip's parcels, looked in the direction of his nod and saw the shattered, barely recognizable remains of a jeep several thousand feet down the precipitous cliff just beyond the shoulder of the road. From this, and many other similar episodes, Rip developed the reputation of "a guy who doesn't screw around!"

# 2

## RENO, NEVADA

To every Pacific Fleet carrier pilot Reno, Nevada, is almost like a second home. It is the closest major city to Naval Air Station Fallon, Nevada, the home of the Naval Strike Warfare Center and a vast complex of training targets and simulated Russian surface-to-air weapons simulator facilities. In 1954 it was a very austere auxiliary air station whose value to Pacific Fleet carrier aviation was its proximity to local training targets. In those days there was no fancy strike warfare center, no sophisticated surface-to-air threat simulators, no tactical air combat training systems; only perfect flying weather and targets just minutes flying time away. Even though our squadron pilots were assigned open-bay barracks buildings with double-decked metal bunks and communal showers, we loved it! The reasons were several. First, and foremost, we all loved shooting the guns at aerial targets . . . and we got to do it two or sometimes three times a day.

Then there was the camaraderie. We were all a captive audience to the excitement of the bottom line of any combat organization, the live-firing of weapons in simulated combat. We were all out in the middle of nowhere . . . the high desert of northwestern Nevada. There were no wives, girlfriends or dependents waiting for us to show up 20 minutes after quitting time. We lived the simple life and lived in our flight suits. Every gunnery flight carried with it the standard wager. The lowest-scoring pilot bought a round of beer for his three flying mates. The debt was payable that night either in the officer's club or downtown in one of Fallon's many gambling casinos. On a Friday night, if someone had a car, those debts were often paid in one of Reno's gambling houses 60 miles to the southwest. The high desert air was clear, dry and crisp. The summer days were hot and the nights cool. Winter days were bearable due to the warming effect of the blazing sun. Winter nights, however, could be cruelly cold depending upon the wind. We all loved the gunnery deployments. They were generally planned to be at least three weeks in length because, in part, it took two weeks to sharpen up one's shooting skills to the point where we could consistently hit well.

The message came in to Air Group NINETEEN's office early in September 1954. It invited participation by Pacific Fleet aviation units to provide static display aircraft for an annual open house celebration. The message was sent from Stead Air Force Base near Reno, and was addressed to a host of Navy, Air Force and Marine Corps aviation organizations on the West Coast. Although it was a poorly written message, vague and ambiguous, there was no doubt about the time and day of the event. When I saw the message on the VF-191 message board, I went immediately to the squadron operations officer and volunteered. I offered to fly an airplane to the event, stand alongside it during the two-day weekend event, look square jawed and steely-eyed in my flight suit to answer questions. Furthermore, I promised to be polite, the very epitome of the dedicated naval officer, and to return the precious airplane on Sunday afternoon unscathed. Although there was a standing policy not to let a junior officer go on a cross-country flight alone and unsupervised until he had proven his reliability (and I had not done so) an exception was made to the rule and my request was granted. I was delirious with delight!

When Friday came I was genuinely excited. The air show officials wanted all static display aircraft delivered no later than 2:00 p.m. Friday so that they could be towed to the display area and tied down for the next day's exhibit to the general public. I was anticipating an easy, one-hour flight to Stead as I climbed the boarding ladder into that newly washed and waxed Cougar. What had not been clear in the message was that Stead AFB, acting as the military surrogate for Reno Municipal Airport for the first time, was in fact inviting military participation in the annual Reno Air Races. Oblivious to this subtlety, I was cruising my Cougar through beautiful clear skies, across the snow-capped Sierra Nevada Mountain range at 45,000 feet leaving a long, white, flossy vapor trail; a very happy, if somewhat ignorant young pilot.

As I approached the Reno area I dialed in Stead Tower, identified myself as a static display airplane, and requested landing instructions. I had already begun a long, steep descent when the Stead Tower operator burst my bubble.

"Navy Jet Alpha Bravo One Zero Seven, this is Stead Tower. There's no open house here this weekend. Where you're supposed to land is over at Reno Municipal Airport, just 15 miles southwest of us, over."

I was dumbfounded, and looked down at Reno Municipal Airport off my left wing. "Jesus," I thought to myself, "It looks awfully small." I thanked Stead Tower and leveled the airplane off with a burst of power while I fished frantically for my

Radio Facilities publication (RADFACS). Looking up Reno Municipal in the magazine-sized publication, I was horrified to note that the longest runway was only 6,600 feet long. That would have been only 600 feet longer than the minimum 6,000 feet specified in the squadron standard operating procedures at sea level. Unfortunately, Reno was not at sea level. It was one mile high! This meant that stopping distance would be 25 percent greater than at sea level.

Airplanes land at certain indicated airspeeds (or angles of attack for newer models). True airspeed increases at the rate of two percent per 1,000 feet increase in field elevation. Thus, if I flew my Cougar to a landing at the prescribed 135 knots indicated air speed, my true airspeed at Reno Municipal would be about twelve percent higher, or 162 knots. Since the energy required to be absorbed by my braking technique was a direct function of the mass times the square of the velocity (my mind was racing through the mental calculations at this moment), I knew it would take at least 25 percent more runway in which to get stopped. I was certain I could not do it unless I flew an absolutely perfect landing pattern. What a dilemma!

The last words of the Skipper rang in my head. "Take it up there, land it, answer questions and bring it back. No screwing around . . . understand?" I could feel the huge knot beginning to take shape in my stomach. I was in trouble and I knew it! My mental calculations were shattered by another crushing blow.

"Navy Jet, Alpha Bravo One Zero Seven, this is Stead Tower, be advised that there is an existing NOTAM (Notice to Airmen) out which indicates that there is only 4,700 feet of usable runway on the long runway due to construction on the approach end. Ensure that you land 1,000 feet long, over." My mouth was suddenly so dry that my voice croaked a little when I acknowledged that transmission. I hadn't checked the NOTAMS for Reno because I never intended landing there in the first place! The sweep second hand on the eight day clock on the instrument panel marched around the dial while I pondered my alternatives. Prudence would dictate that I alter my flight plan, divert to either Stead or Fallon (sixty miles east), gas up and go home with my tail between my legs. How could I ever explain not making a simple telephone call to Stead which would have cleared everything up while I as still on the ground at Moffett?

I was feeling cornered and grim when, all of a sudden, my ego kicked in. Wait a minute, I reminded myself. I am a fully qualified Navy carrier pilot. I can land my Cougar on the deck of a carrier. Why not burn down my fuel to just a few hundred pounds remaining, fly a carrier landing pattern and make a full-cut carrier landing at a few knots slower than recommended? If I do a perfect job and grind down the skag (a steel plate on a hydraulic cylinder which extended with the wheels down to prevent the tailpipe from contacting the runway), I can get this thing stopped before I run out of runway.

My alter ego kicked in in rebuttal. "Listen, you dummy," it told me. "If you run off the end of the runway, blow both tires and have to get lifted back onto the concrete with a crane in front of 120,000 people, the Skipper is going to have your ass!" My ego won over my alter ego, and with my heart in my mouth I dialed in Reno Tower frequency.

"Reno Tower, the is Navy Jet Alpha Bravo One Zero Seven. I am overhead at 20,000 feet and will have to delay my landing while I burn down about 2,000 pounds of fuel for about 15 minutes, over."

"Navy Jet Alpha Bravo One Zero Seven, this is Reno Tower. Why do y'all have to burn down fuel? Over." I explained the reasons and the country voice with the cowboy drawl came right back at me. "Navy Jet Alpha Bravo One Zero Seven, this is Reno Tower. Y'all can burn down fuel by putting on a whiz bang air show for all these fine folks down here. I'll just close the field for you right now. We've got over a 100,000 people down here who'd love to see what a brand-new, supersonic Navy jet can do, over."

That did it!

I don't know what came over me, but I told him I was on the way down. It should be noted, at this juncture, that the F9F-6 Cougar was capable of supersonic flight; but only in a vertical dive at full throttle entered at 45,000 feet. The airplane would pay a brief and transitory visit to the world of supersonic flight passing through about 35,000 feet and stay there, at about mach1.05 until it reached about 32,000 feet when it would decelerate back to the subsonic regime.

As I rammed the throttle to the firewall and made an approach down the show runway, the white fangs came out and all common sense left me for the next 15 minutes while I "beat up" the field. The first pass was at 580 knots and I started a loop directly in front of the main reviewing stand from an altitude of about 50 feet. I came out of the loop and back before the crowd at 550 and, pulling up into the vertical, commenced a series of eight aileron rolls going straight up until I ran out of airspeed somewhere about 15,000 feet. And so it went for another 10 minutes. There were a few Cuban Eights, and Immelman turn, a squirrel cage and some other maneuvers I can't name. It was a totally unauthorized, foolish, risky and asinine performance, but I loved every second of it! I realized, when the low fuel warning light illuminated, that this was the first airshow of my career . . . and, if the Skipper ever found out about it, the last! One more low pass and I faced the acid test, the landing on that incredibly short runway.

But the decision process was over. With only 400 pounds of fuel left, I couldn't even go to Fallon. There was no other alternative but to land at Reno . . . Stead was closed.

I couldn't have screwed this flight up any worse, I thought as I entered the "break" and turned downwind for a final landing. The thought passed through my mind that this might very well be the "final landing" of my career. Instead of an 800 foot downwind leg, I dropped it down to 200 feet. Instead of a 135-knot landing speed, I slowed it down to 130 knots. Coming out of the final turn, I decided that the first 100 feet of concrete was the deck of an aircraft carrier. There, on the unusable first 1,000 feet of runway, were the bulldozers and earth-moving equipment involved in the runway repair. I was going to simulate a carrier landing, taking a full-power cut from an altitude of about 100 feet and "lay it in the spaghetti." I knew I could do that as well as anybody. The airplane came down like a ton of bricks and I pulled up the nose plenty high, keeping the nose wheel well off the concrete to get as much aerodynamic braking as possible. I could feel the skag plate make contact with the runway, and, knew it would be ground down to nothing in a matter of seconds. Well, I thought. That's what it was for!

After about 2,000 feet of roll out I could see the end of the runway coming at me like a freight train. The airspeed still read 90 knots, but I couldn't wait any longer and, dropping the nose wheel onto the concrete, I "got on the binders" and said a silent prayer for the tires to hold together. By now, the end of the runway was clearly in view and I was sure the Cougar wasn't going to stop in time. The concrete ended abruptly and after about an 18 inch drop off, there was nothing but dirt and the ground sloped away rather steeply to what looked like a school building. With about 75 feet to go, I locked the brakes and the tires began to skid . . . the precursor to blowing. For some reason, known only to God, the tires did not blow . . . and the Cougar came to a gentle halt. There was about 10 feet more of concrete left. I had done it! The sweat just ran off my face.

The Cougar was halfway down the taxiway to the performer's parking area before my heart rate and breathing returned to normal. The ground control operator directed me to a parking place directly in front of the reviewing stand where the air show coordinator was located. When I shut down the engine, I could hear the announcer giving the audience my name and making some rather grandiose statements as to what the crowd had just seen . . . including supersonic passes and other nonsense.

As I climbed down the boarding ladder, I felt someone grab my flight boot. Looking over my shoulder I saw that a white Cadillac convertible had pulled up close to the airplane. It was full of show girls. One of them was guiding my foot onto the seat cushion of the back seat. I stepped in and got a hug and a kiss from each of the show girls as the Cadillac paraded me up and back in front of the grandstands. The announcer asked the crowd to give me a round of applause for my performance. After that I was whisked downtown to a complimentary room at Harrah's Club. On top of the dresser in my room was a huge pile of one-dollar gambling chips. I had a truly wonderful evening in Reno that night.

The next day I showed up alongside my airplane to answer questions and to watch the airshow, which was truly spectacular. In between events, an Air Force T-33 came in and made three landing attempts, waving off each time after rolling a short distance down the runway. I never found out where he went . . . probably Fallon.

One part of the air show featured a simulated attack by VA-192, a Panther squadron in the air group. They came in from Fallon, staged their attack and returned to Fallon.

Sunday afternoon's flight back to Moffett was uneventful. On Monday morning I entered the ready room and gave a cheerful, "Hello, good morning" to the executive officer. He just glowered at me, and I knew immediately that I was in trouble. The duty officer told me the C.O. wanted to see me. I felt an ominous chill when I went in and shut the door behind me when directed to do so by the Skipper, "Butch" Voris. He didn't waste any time getting to the point.

"Paul, what did I tell you to do at Stead Air Force Base last Friday?" he asked.

"To land there, stand by the airplane during the show and bring it back here safe and sound." I answered, dreading what I knew was coming next.

"Cy Whitehead told me you put on quite an air show!" he said, now looking me straight in the eye. I dreaded those cold blue eyes.

"Sir, I know I wasn't supposed to, but he invited me," I started out lamely. I now knew that my flying career was over and I felt truly miserable about my own stupidity. Cy Whitehead was the C.O. of VA-192. He must have been in the stands when I put on my show.

"The X.O. recommended that I throw you 'in hack' (the equivalent of house arrest), Voris continued. "But, I decided not to do that. Do you want to know why?" he asked. I nodded dumbly.

"The reason why I'm not throwing you in hack for disobeying me is that I can't understand how you ever got that airplane stopped without running off the end of the runway at Reno. I've got 5,000 hours of flight time and I wouldn't even dream of trying it. You must have done everything perfectly and gotten away with it. But, here's a warning. If you ever disobey me again, you're through. Is that clear?"

He never raised his voice, but I knew from that cold stare and the quiet tone that he was deadly serious. As I walked out of his office I let out the lungful of air I had been holding and made a solemn vow never to fool around again . . . never!

# 3

## GUNNERY DEPLOYMENT

The world-famous Satan's Kittens deployed to Naval Auxiliary Air Station, Fallon, Nevada, from 15 May through 7 June 1954, along with the rest of Air Group NINETEEN. Our training consisted of two flights per day per pilot in aerial gunnery and air-to-ground strafing. We all loved to fire those guns. The first detent on the trigger (located on the control stick) started the gun cameras whirring. A full squeeze of the trigger (through the second detent) started all four guns firing. The chugging of four 20mm cannons really made the airplane shake as it exhausted a deadly stream of steel. There were three incidents which marred what would otherwise have been an extremely enjoyable flying interlude.

The first incident was the shooting down of the tow plane. A utility squadron, VU-7, was providing aerial banner towing services for VF-191. One of the tow pilots, Lieutenant (junior grade) Don Heriot was towing a banner at 20,000 feet for Al Franier's four-plane division. One of the nugget pilots in Al's division, Lieutenant (junior grade) Walt Wray, was having difficulty mastering the pattern. As a consequence, he was not getting any hits in the banner. A great deal of effort had been expended by his section leader, division leader and even a couple of the squadron's seasoned sharpshooters trying to get him to improve his skills. Aerial gunnery was, and still is, one of the most demanding missions for a fighter pilot to master. Walt tried hard, but he was not one of the more skillful nugget pilots in the squadron.

The protection of the tow plane from stray bullets was enhanced, if not ensured, by three hard and fast safety rules. First, the firing plane, during the firing portion of the run, must always be descending – never firing up at the banner. Since the banner sagged 100-200 feet below the tow plane, the rule was a fairly good guarantee of the tow pilot's safety.

The second rule required the shooting pilot to maintain a hard turn (good "angle off" the path of the banner) during his firing burst. This angle off must never be less than 15 degrees, thus causing the stream of bullets to carry past the banner and away from the tow plane. The third rule was intended to enhance the safety of both the shooter and the tow pilot. It was the minimum-range rule which required the shooter to break off his run at a firing range of no less than 800 feet. Many a violator of that rule has come home with the 10 foot steel tow bar (from which the leading edge of the banner is suspended) buried deep in the leading edge of his wing . . . not a pretty sight! The minimum-range rule also provided a modicum of safety for the tow pilot simply because the likelihood of a 20mm round glancing off the tow bar and carrying all the way up the tow cable to the tow plane was extremely remote.

But all these rules notwithstanding, Walt Wray's gunnery runs this particular day were just as futile as his earlier tries to "put a few beans in the rag." His firing passes began to get progressively flatter. The angle-off on his runs became smaller and smaller, resulting in what is called a "sucked" run. And, compounding the felony, he began firing closer and closer in. The tow plane was approaching the end of the gunnery range and termination of live firing runs was imminent. Walt's last firing run was nearly fatal for Don Heriot.

Don was a likeable, easy-going young man whose approach to flying was casual in the extreme. Whenever it was my turn to escort the banner back to the drop zone, I always rendezvoused on the tow plane, gave him a wave, then slid slowly back down the tow cable until I stabilized abeam of the banner. The escort fulfilled two functions. The first, a safety function, was by its presence to keep other pilots from flying into the cable (which was nearly invisible) or into the banner which was also very hard to see. The second function of the escort was to see where the banner landed if it fell off the tow plane enroute back to the field. We often went to great efforts to retrieve lost banners . . . especially if there were lots of holes or if the bet was high. When the tow plane returned to the drop zone, the escort called the drop as soon as he could see that the banner was inside the fence.

Never in all of the escort missions I flew on Don Heriot did I ever see him with his oxygen mask on (a violation). More often than not, he was smoking a cigarette (another violation) or munching on a package of Lorna Doones. I even caught him reading a magazine once. I remember one day the Skipper was looking out the second-story ready room window as Don taxied in and parked his tow plane. Don came booming in much too fast; both elbows casually draped over the canopy rails with his flight suit sleeves rolled up all the way to his elbows (another violation). He was going so fast that the taxi director, after giving him the left turn signal into the parking spot, had to give him an immediate emergency stop signal.

Don locked up both brakes and came to a halt, leaving two black skid marks 10 feet long on the concrete.

"That young man will never fly one of my airplanes," the Skipper grated. This was in reference to an informal request Don had made to let him check out in one of our red-hot, swept-wing fleet fighters. The tow planes were tired out old Panthers. The incident sticks in my mind. It was vintage Heriot.

But Walt Wray's last run combined violations of all three rules in one deadly run. Although the accident investigation board never proved exactly from whence it came, it did conclude that an object exactly 20 millimeters in diameter entered the engine through the tailpipe and punctured one of its combustion chambers. The resultant fire spread rapidly and Don Heriot ejected from the burning tow plane. He later testified that he observed "sucked" and uphill runs by the number four man in the flight. It was immediately after the worst of those runs that the fire warning light on his instrument panel illuminated. When the flight leader informed him, moments later, that he was on fire, the "nylon let-down" was the only appropriate course of action.

An extremely concerned Skipper of the utility squadron decided to fly up to Fallon to talk with "Butch" Voris, the Skipper of the now more world-famous Satan's Kittens. He came up in one of his tow planes from Moffett Field and towed a mission himself to see how well we were adhering to safety rules. He chose a JD-1 (a Navy tow plane version of the famous World War II vintage Douglas A-26 bomber). Although propeller driven, the JD-1 had an advantage which the faster Panthers didn't have. It could reel in a cable in flight and attach a new banner in case one was shot off by a stray bullet. The JD-1 tow mission was uneventful except for the fact there was a sudden rise in cylinder head temperature in the port engine. Exercising proper precautions, the VU squadron Skipper brought the banner back to Fallon, dropped it and then landed and asked his crew chief to check out his port engine.

It was during this meeting with Voris that the crew chief burst in with ominous news. It seems that the sudden increase in cylinder head temperature was the result of damage caused by an object which had punched a hole in the port engine. THE HOLE WAS EXACTLY 20 MILLIMETERS IN DIAMETER! I recall that it was the executive officer's division that was firing on that tow mission. We were all walking on egg shells for the next few days.

The third incident, just as unnecessary as the first two, was a tragic one which resulted in the death of my good friend, Lieutenant (junior grade) Don Krause. Don had lived in the room next to mine at the BOQ in Alameda while we both awaited the return of VF-191 from WestPac. Don was quiet, bashful and extremely warm hearted. He was also very religious. I interrupted him more than once in his room at night saying his rosary beads. He was a daily communicant during those spring months and we saw a great deal of each other. I never heard him take the Lord's name in vain, and I never heard him speak ill of anyone. Don was of medium height with a trim build, a cap of tight blond curly hair and a broad grin that lit up his face, revealing a row of perfect teeth. But, it was a self-conscious kind of grin. Don and I were both nuggets, newly reporting to VF-191 . . . and we were friends.

It was near the end of our third week at Fallon and Don was scheduled for a strafing mission on one of the targets just northeast of the field. A second-tour pilot, Lieutenant Willie Haff, was also in the flight and a witness to Don's fiery death.

The strafing banner was a polyurethane rectangle (two aerial banners sewed together), 12 feet high by 30 feet wide, and oriented to the ground at an angle of 30 degrees from the vertical. A race track pattern was flown with the run-in line oriented at 90 degrees to the banner. Pattern altitude was 4,000 feet and the briefed speed in the pattern was 350 knots. Adherence to speeds and altitudes served to enhance safety in the pattern and also served to preclude "bunching up." In a properly flown air-to-ground weapons pattern, the airplanes were evenly spaced around it and able to observe one another throughout. The roll-in point was on the extended run-in line. The pilot made a 90 degree, nose-down turn toward the target establishing about a 20 degree glide angle and adjusting the throttle to achieve a speed of 450 knots at the moment of firing. At 300 feet above the terrain, with the proper sight angle on the target, the pilot had time for a quick one-and-one-half second firing burst before he had to initiate his pull-out. To avoid "walking" his rounds through the target, the good strafing pilot eased some of the back stick pressure that he was holding during the firing burst. This had the effect of steepening the glide angle slightly and focussing the impact point of the bullet stream. If the gunsight lead angle was right, the tracking smooth, the airspeed correct and the firing altitude exactly right, that compressed impact point would be on the target and the pilot would score many hits. If, however, the pilot succumbed to the temptation of watching his bullets hit the ground, so also would his airplane. There simply wasn't any margin for error. The ground was too close, and the plane was traveling too fast for any other alternative to be available. There was no time to watch the bullets hit. An immediate pull-out was mandatory because ground impact was only seconds away. The phenomenon known as "target fixation" has killed more air-to-ground pilots than any other aspect of tactical weapons delivery. As a consequence, its dangers are assiduously briefed before every mission.

No one will ever know why Don Krause did it, but he initiated a late pull-out. Willie Haff was rolling in from a position dead astern and watched the whole sequence of events beginning with the moment the tailpipe of Don's airplane made contact with the ground. It left a shallow groove in the sandy soil just beyond the target. The groove ran for about 200 feet then stopped abruptly as the airplane became airborne again. But, the impact had been enough to tear the roaring jet engine loose from its heavy mountings sending its white hot flames into an inferno of ruptured fuel tanks and spurting fuel lines. In the few micro-seconds which followed the cloud of dust left by the tailpipe in the sand, Willie knew that Don Krause was dead. There followed a violent explosion, an orange-and-black fire-ball and pieces of debris spattering in the sand in small dust spurts for a quarter of a mile beyond the target. There weren't very many big pieces left. The target crew found Don's body surprisingly intact. There weren't any signs of the severe internal injuries which he had suffered. He had been blown free from his ejection seat and was lying there on the sand almost peacefully. Death had been instantaneous. I don't believe Don ever felt a thing!

Lieutenant (junior grade) Bob Patterson and I volunteered to drive Don's car and personal effects back to his parent's home in Oshkosh, Wisconsin. We were dispatched immediately to Moffett Field where we inventoried his belongings, packed them in his car and headed east. The trip was pleasant enough until we neared the terrible, mind-numbing experience of meeting his family. Don, a bachelor, was one of a large family of devout Catholics who knew absolutely nothing about the Navy, or about carrier aviation beyond what Don had told them in his letters home.

We had made up our minds that, under no circumstances, would we accept the hospitality of their home. It just wouldn't be right to stay in the home at such a difficult time in their lives. However, we hadn't counted on what awaited us there. Don's Mom insisted on our staying. One look in her eyes, and I caved in. There simply wasn't any way to refuse. It was obvious the whole family really wanted us to stay. Bob and I would much rather have put up at a nearby hotel where we could call up room service and order two cold beers.

So we stayed there, and were enriched for the experience. But it was a little like a stay in a monastery. Life in the Krause household was simple and the family was devout. While we waited for the funeral arrangements to be completed, we were invited to attend rosary sessions, meet the clergy who all knew Don as an altar boy and member of the church choir, and other parishioners who knew him as an active member of the congregation.

The countless questions of how and why, and did he suffer, were extremely difficult to answer. Strangely, I can recall very little about those poignant hours in the Krause household deep in the center of America's heartland. Of course, we kept the squadron informed on the progress of our mission with periodic telephone calls. When the funeral arrangements were finalized, we told the squadron and they arranged for our retrieval. We were not on official orders. Everything was being done on the cuff. There was no money for airline tickets.

Lieutenant "Monty" Montgomery had some previous experience flying the Douglas AD-5 Skyraider. He borrowed a four-place, night fighter version of the airplane from the composite squadron at Moffett (VC-3) and flew it to Madison, Wisconsin, to bring us home. The six or seven days which elapsed between Don's death and our return to Moffett Field will always retain a dreamlike texture in my memory.

There was one more casualty of that deployment to Fallon which I recount with trepidation, because the casualty was our executive officer's ego! One of the kindest, gentlest and most warm-hearted individuals I have ever known, Lieutenant Commander Tiny Graning had a problem, as do many of us, with weight control. A huge-framed man, he stood 6 feet 5 inches tall and must have weighed 280 pounds. He was always on a diet, but succumbed, as many of us do, to a sweet tooth and the consequences of gracious living.

But this time it was going to be different, he announced to the whole squadron. Yes, by God, a three-week gunnery detachment to Fallon, Nevada, with its intense flying schedule, living in our flight suits and its austere regimen was the perfect place really get serious about dieting and "knock off 30 pounds or so." And, to his credit, he did get serious. He chose salads at the officer's club mess, passed up the mashed potatoes and, in general, tried hard to eat smaller portions of everything put before him. Unfortunately, all of us were made to suffer along with "Tiny."

The continual and excruciating descriptions of bodily deprivation; and his running commentary on the paltry portions and tastelessness of "rabbit food" began to take their toll on all of us. For two members of his flight division, Lieutenant Jack Finney and Lieutenant (junior grade) "Dirt" Pringle this non-stop soliloquy became too much to bear. Even though Tiny actually began to make progress in his diet, and had already shed a few pounds, his bragging about loose pants became the last straw!

Our two young conspirators decided to put an end to it by cruelly convincing Tiny that he was, in fact, gaining weight!

Since we suited up for flight in the squadron ready room in full view of all the other pilots, the weapon of choice for this cruel prank was "Tiny's" anti-g suit. The g suit, as it had come to be called, was a nylon Pants-like contraption that contained rubber bladders located to apply powerful pressure to the lower abdomen, calves and thighs. Three zippers, one at the waist and one on the inner side of each leg (from crotch to ankles) were used to put the suit on and take it off.

In theory, under high g (acceleration) loads, high pressure engine bleed air was vented via a hose to the suit where it inflated the bladders. The inflated bladders prevented blood from pooling in the abdomen and legs, thus keeping an adequate supply of blood, and therefore oxygen, flowing through the brain. A properly fitted g suit was supposed to prevent black-outs under high g loads. In actual fact, a well adjusted g suit probably raised a pilot's g (black-out) threshold about one and a half to two gs.

But to be properly fitted, it took a parachute rigger about 20 minutes to tighten about half a dozen corset-like nylon line adjustment points. The rigger would lace up the criss-crossed nylon lines through loops, pulling each one tight as a drum. Then he carefully snipped off the excess line and, with a cigarette lighter, melted the bitter ends of the knotted lines so that the fibers wouldn't unravel. Once done, the g suit was tailored for only one pilot in the world.

Each night, after flight operations were over and after Tiny had left the hangar headed for the BOQ, our giggling co-conspirators laboriously untied each of the many knots on Tiny's g suit, snipped off two inches from each line, retied the knots, and, with a cigarette lighter, melted the bitter ends of each lace. The net effect of all this effort was a gradual tightening of the g suit. It soon became a squadron-wide conspiracy and a tribute to all the pilots who showed monumental restraint as, twice each day, Tiny suited up for flight. Each day he struggled harder and harder to pull up those damnably protesting zippers. Nobody laughed or stared openly. Each of us was inwardly in convulsions, but the poker faces that we all maintained bespoke great inner strength, willpower, and self-control.

I never learned when or how Tiny finally caught on to the conspiracy, for he never told us. But, he took the joke with his usual good grace and tremendous sense of humor. I have always suspected, however, that Tiny's ego was badly bruised for some time thereafter. In retrospect, it was a cruel joke. But then, fighter pilots have never been great respecters of each other's feelings!

# 4

---

## "NO GUTS, NO GLORY"

W hen U.S.S. ORISKANY arrived in Yokosuka, Japan, in the fall of 1955, she was badly in need of a new flight deck. The decision had been made before her departure from Alameda to have the work done upon her arrival in Japan. The savings from replacing the teakwood deck in the Far East versus in a West Coast ship yard promised to be substantial.

Yokosuka, a major seaport east of Tokyo, boasted magnificent shipbuilding and repair facilities. During World War II the port fielded major elements of the Imperial Japanese Fleet such as the battleship Yamato and many of its aircraft carriers. About 30 miles north of Yokosuka was a major air facility named Atsugi, which also supported the Japanese fleet, and at which most of the kamikaze pilots were trained. In 1955, when ORISKANY arrived, Naval Air Station, Atsugi was the major support facility for aircraft from U.S. carriers visiting Yokosuka. The air station was located near a small town named Sagami Otsuka. As soon as ORISKANY was within flying distance of Atsugi, Air Group NINETEEN'S airplanes began flying off to free up the flight deck in preparation for its replacement. However, for the three weeks that it would take for the swarm of shipyard workers to replace Oriskany's flight deck, there was not enough ramp space for all of the air group's airplanes. As a consequence, it was decided to locate VF-191 at nearby Johnson Air Force Base. We shared a parking ramp with a couple of U.S. Air Force fighter squadrons, and agreed to conduct tactics training with them and to pick up a portion of their air defense ready-alert commitments.

The Air Force squadrons were all fired up over the tactics idea. Air Force Captain "Boots" Blesse, a Korean MiG ace, had just spent two weeks on a road show in the Far East with Air Force squadrons teaching the brand new Air Force

---

*USS Oriskany (CV-34) on December 6, 1950 - straight deck configuration. (Official U.S. Navy photograph)*

jet fighter tactics doctrine. It featured a formation called the "fluid four," and Captain Blesse named his road show, "No Guts, No Glory."

The U.S. Navy, on the other hand, was still in the process of developing its jet fighter tactics doctrine based on its newly acquired Korean combat experience. But their Korean combat experience had been acquired with propeller-driven airplanes like the AD Skyraider and first-generation jet fighters like the Grumman F9F-2 Panthers. The newer swept-wing Navy jet fighters like the North American FJ-3 Fury and the Grumman F9F-6 Cougars never got into the Korean combat arena. They arrived in the fleet at the very end of the Korean conflict. In fact, one fighter squadron each in Air Groups NINETEEN and FIFTEEN were the first swept-wing fighter squadrons in the Pacific Fleet. Satan's Kittens, having just completed a 7th Fleet deployment in the summer of 1954 had been charged by higher authority with developing a fighter tactical doctrine.

In our work-up training for this deployment, we had developed a tactical doctrine formed around the four-plane division as the basic fighting element. We called our tactical formation the "fingertip" formation, with each of the two-plane sections providing mutual support for the other in free-play maneuvering to defeat a multi-plane formation of hostile aircraft.

The formation, like the tips of one's fingers, consists of two pairs (sections) of airplanes flying abeam one another (co-altitude), and separated by a distance which varied with altitude. The distance which separated the two section leaders was determined by the arc of rearward visibility behind each section and permitted each section to "clear" the other's tail from attack by the current Soviet air-to-air weaponry. The two sections were mutually supportive and the tactics were largely the same as Captain Blesse's once the battle was joined. As we applied these tactics in our daily air combat maneuvering training, we learned that division integrity tended to break down after the first minute or so. Section integrity was only maintained by adamant insistence on the basic "jungle" rule – a wingman never abandoned his section leader. Single airplanes in a multi-plane engagement were "dead meat." The section leader had to be able to provide mutual support to his division leader by being virtually assured that his own wingman would keep his own tail clear.

So it was with great anticipation that both the Navy and Air Force squadrons looked forward to trying out their new found tactical doctrine against one another . . . and a dissimilar airplane . . . to add reality. I was the number-four man in the Skipper's division. Our new C.O., Butch Voris, was a legend in his own right, having been combat tested in World War II, Korea, and having been the founder of the now-famous "Blue Angels."

In the late 1940s Butch had founded the first Blue Angel team, when they were doing their aerobatic work in the legendary Bearcat fighter. When the Korean conflict broke out, the Blues were disbanded and their members distributed among several Pacific Fleet squadrons including Satan's Kittens. After the Korean conflict ended the Blue Angels were re-constituted, and Butch was called upon to start them up again. He went from that tour of duty to commanding officer of the Satan's Kittens.

This time, however, the Blues were flying jet-powered aircraft. Because of their low thrust-to-weight ratio (underpowered), there was a substantial time lag between aerial maneuvers over center stage. Two solo airplanes were added to the act to keep the audience's attention while the four-plane team was building up kinetic energy for their next maneuver. Butch taught us (his own division) all of the basic Blue Angel aerial maneuvers when time permitted. He was, without a doubt, the smoothest flight leader I have ever known!

But by far, his greatest asset in the aerial combat arena was his tremendous eyesight. Of all of our four-versus-four engagements with the Air Force at Johnson Air Force Base, we never lost a single battle. The Air Force F-86Fs had many advantages over our Cougars. They were lighter and therefore more agile, with a greater rate of climb and better acceleration. The only advantage the Cougar held was that it could pull more g than the F-86s. The Sabres were limited to seven gs while the Cougars could pull nine. Of course, both planes were so underpowered that either a seven- or a nine-g turning fight was a downhill maneuver.

The key element in our division's success was the Skipper's eyesight. He always saw them before they saw us, so we maneuvered early in each engagement to be in a position of distinct advantage before they ever caught sight of us. We began every engagement with an offensive attack. The Skipper's technique after such a start was to keep our speed high enough so that we could take full advantage of our two-g advantage. During one particular engagement with the division of F-86s lead by the Skipper of the Air Force squadron, he cracked a wing spar by pulling excessive gs. During the debrief, he asked Butch what he ought to be doing in order to win the next engagement. With a twinkle in his eye Butch Voris told him they ought to paint their airplanes another color. He was only half-kidding. The Air Force was very proud of the beautiful silver color of their fighters . . . they polished them daily . . . and were not about to change the color scheme. But it was true; the silver planes were much easier to see than our two-tone blue ones.

I didn't keep score as to how successful our other divisions were against our Air Force opponents. I suspect we all came out about even. Nor did we prove that the "fingertip" formation was any more superior to their "fluid-four" formation. I suspect that we both learned that our own tactical formation was best suited to the particular characteristics of our own airplanes. But one lesson came through all of our engagements, loud and clear. The best-trained division always won!

It was not surprising to any of us that Lieutenant "Rip" Rohrer provided one of the lighter moments of our three weeks tour of exchange duty with the Air Force. Rip was the section leader in the third division, and had really "tied one on" while on liberty the night before one particular training-day's operations. He looked like death warmed over when he showed up in the briefing room for the early launch. The four Air Force pilots in the briefing room looked askance at him, obviously wondering how in the world he was ever going to even find his own airplane much less fly it in simulated combat. For whatever reason, his division leader, Lieutenant Commander Al Franier, chose to ignore Rip's terrible condition when the flight briefing was over.

The eight pilots walked to their airplanes parked alongside one another on the ramp, with Rip trying mightily to keep a steady course across the concrete parking ramp. When he went to climb up the boarding ladder of his airplane, his foot missed the lower step and he fell forward heavily onto the concrete apron, his head striking the lower airplane step on the way down. Al Franier was manning the adjacent airplane and was watching when Rip fell. He ran over to Rip's inert form, as did two of the Air Force pilots. When they reached him, Rip had already struggled to his feet completely oblivious to the ugly looking gash in the side of his face. Blood gushed onto Rip's flight suit, but he didn't notice it and started back up the ladder. The Air Force division leader stood there looking aghast! He and Al Franier exchanged knowing glances and Al made his now-famous observation. "Rip," he announced in a loud voice, "We are going to have to scratch this mission because of bad weather."

Rip stopped, one foot on the bottom step of the ladder, and looked up at the sky with bleary, red-rimmed eyes. There wasn't a cloud to be seen anywhere, just endless blue – the visibility must have been 70 miles.

"Yeah, I guess you're right," he slurred and started back in a sinuating path toward the hangar.

# 5

## "LIBERTY, BUT NO BOATS"

Lieutenant Brooke Montgomery was quite an unusual naval officer. As a second-tour pilot, he was a leader of one of the squadron's four-plane divisions. Of medium height and build, he had a ruddy complexion, clear blue eyes, and sported a butch haircut turning prematurely gray from its natural darkbrown hue. Brooke was meticulous in dress, manners and speech. He was, in a word, a patrician in a squadron of spirited and often raucous fighter pilots.

But, on the return trip of U.S.S. ORISKANY to the States in October 1955, he became something of a legend in that aircraft carrier's history book. The ship was on its way back from a deployment to the Western Pacific and had to make an administrative stop in Buckner Bay, an anchorage on the southern coast of Okinawa. Her hangar deck and flight decks were jam-packed with all of the air group's aircraft, the ship's liberty boats, aircraft ground support equipment and assorted impedimenta which had been located at a beach detachment at Naval Air Station Atsugi, Japan, for the major portion of the deployment. There would be no flying until ORISKANY made preparations for a fly-off of the air group's aircraft a few hundred miles west of the California coast. That exciting event was still a full month away.

The ship's executive officer made an unforgivable blunder in basic leadership. He announced over the ship's public address system (known as the 1MC) the fact that we would be remaining at anchor for only 18 hours. Because of the time it would take to take the liberty boats out from their locked location in Hangar Bay 3 and then to stow them again, the decision had been made not to do so. However, he added with a dead-pan voice, liberty was available for anyone who managed to get ashore by any other legitimate means. I listened to this idiotic announcement with shocked disbelief. What the XO meant to say (I knew), was that if there were room in the several small boats which were shuttling back and forth to the beach on administrative errands (or on the helicopters), those who could manage to wangle a seat could go on liberty.

What he had actually repeated, unintentionally, was a version of the stale old navy bromide about the captain who announces that there would be "Liberty but no boats." It didn't matter much to me, but I knew that there must have been loud hoots of derision from the mess decks and crew quarters all over the ship at this gross faux pas. There was very little to see or do ashore in this relatively uninhabited part of Okinawa. Nevertheless, such a gratuitous announcement was sheer stupidity and bordered on almost a lunatic lack of consideration for crew morale.

Within a few minutes after this Captain Queeg-like announcement, Monty stepped into the junior officer's bunkroom that I shared with five other officers. "Paul, may I borrow your scuba gear?" he asked. Right away I got a funny feeling. Self-contained underwater breathing apparatus (scuba) was a relatively new thing in 1955. It had not as yet caught on as a popular water sport. Commercial equipment was virtually unknown. The only way to get it was the way I did. I bought it used from a retired Underwater Demolition Team sailor who had collapsed a lung using it. There were no diving schools except in the Navy. People who acquired scuba gear had to learn on their own how to use it safely, and the mortality rate of these self-taught neophytes was high. I had scared the hell out of myself a few times while becoming proficient. The diving opportunities on a Western Pacific carrier deployment, however, were tremendous, and I had done some great diving off of the Philippines and Japan.

With some misgivings I gave Monty a 15 minute safety "how to . . ." lecture on scuba diving. Monty gathered up my single compressed air tank, flippers, face plate, rubber mouthpiece, breathing tube, and air regulator and left. He even borrowed my hand spear, for I didn't know what purpose. Over his shoulder he called, "Thanks, I'll return them later," his face a perfect deadpan. I decided to follow him to his stateroom, but kept myself hidden from view. There was now a definite sense of unease and impending trouble.

Monty emerged from his stateroom after only a few minutes wearing a brightly colored, boxer-style bathing suit. He wore the air tank strapped on his back, the face plate on but pushed up onto the top of his head (as I had shown him), the flippers on, the spear in his hand and a determined look on his face. With the awkward gait imposed by the clumsy flippers, Monty made his way aft on the hangar deck towards the quarterdeck. "I can't believe this guy," I muttered to myself as I followed from a discreet distance.

The quarterdeck on a U.S. Naval vessel is a very formal area surrounding the top of the accommodation ladder. It is roped off by decorative lines and guarded

*The following sequence (pages 43-46) shows a typical daytime straight deck carrier landing aboard the USS ORISKANY (CVA-34). These photos were taken from the nose camera of an Air Group Nineteen McDonnell F2H-1P in the Sea of Japan on May 6, 1955 by pilot John Romano. (Official U.S. Navy photographs)*

*The arrow at bottom right points to the dent in the starboard fantail gun tub made by Lt(JG) John Mitchell's F2H-3N. Mitchell walked away unscathed.*

by several sailors decked out in their dress uniforms. A Junior Officer of the Deck (JOOD) carries a telescope more as a sign of authority than its utility. The quarterdeck is the ship's front door, its main entrance for visiting dignitaries. The Boatswain's Mates who guard it had spent many hundreds of hours with marlinspikes weaving the intricate ropes and polishing the teak handrails and shiny brasswork.

Oriskany's quarterdeck watch caught sight of Monty as he stepped around the decorative sign board bearing a color photograph of the ship's captain and executive officer, and the inscription, "WELCOME TO U.S.S. ORISKANY." The watch crew stared in stunned silence as Monty strode purposefully over to the JOOD, stood before him at attention and announced in a loud voice the time honored, "Sir, I have permission to go ashore." He did it so loudly and so formally that the poor JOOD's right hand came halfway up to the brim of his bridge cap in the standard response.

His right hand hesitated as the startled young man stammered, "Very well." Then his hand continued its trip to his cap brim in the time-honored salute. Monty, without hesitation, started laboriously down the accommodation ladder towards a small platform at its base called the boat landing which was suspended by lines three or four feet above the surface of Buckner Bay. All the while, the quarterdeck crew were staring in open-mouthed disbelief at Monty.

When he reached the landing, Monty stopped, put the mouthpiece into his mouth as I had shown him, pulled the face plate over his eyes and nose, and tumbled backward into the water exactly as I had briefed him to do. There was a swirl of foam and seething bubbles obscuring Monty's form. When the foam subsided, Monty was nowhere in sight. All that marked his passing was the steady stream of bubbles coming from below the surface. The stunned JOOD continued to stare, the look of disbelief still on his face as the Chief Boatswain's Mate said to him in a low voice, "Sir, don't you think you should call the executive officer?" I wanted to burst out laughing, but didn't for fear of giving my hiding place away. Besides, I was now genuinely concerned over whether I would ever see my friend or my $600 worth of equipment again.

I watched the continuing stream of bubbles from Monty's regulator and concluded that he must have gone very deep to be out of sight in such crystal-clear water. There was a flurry of activity on the quarterdeck as the executive officer came striding hurriedly across the hangar deck with his Marine orderly maintaining his position "one pace to the rear and one pace left." It was time, I decided, for me to get the hell out of there, and I did so quietly and unobtrusively, keeping airplanes and equipment between me and the gathering throng on the quarterdeck.

Back in my bunkroom I sat on my bunk and waited, in fear and dread, imagining all sorts of horrible things happening to Monty and me. He could, I knew, easily be sucked into one of the ship's evaporator scoops. What a horrible way to die, I thought. As for me, everybody on the ship knew I was the only one who owned scuba gear. It was common knowledge among the Boatswain's Mates who had seen me numerous times going down the same accommodation ladder with my diving gear slung over my shoulder. It would, I knew, be only a matter of time until the executive officer's orderly knocked on my door. After about 15 minutes of such dark thoughts I was astonished to see Monty step into my bunkroom. He was dressed in a neat, impeccable khaki uniform and was carrying my scuba gear still dripping water and leaving a telltale trail to my door.

Monty had an amused look on his face as he described going down under ORISKANY's barnacle-laden keel and up the port side to the other landing used to load garbage onto barges that took it to dumps ashore. As he spoke animatedly about his first scuba adventure, I shuddered. Those huge intakes in the hull which sucked seawater into the evaporators gave me the willies. He could have disappeared into one of them so easily. After he finished his story I told him I would surely catch hell from the executive officer.

"No, you won't," he responded emphatically, "I never broke any rules. After all, he did grant us liberty, didn't he?"

However, true to his honest nature, Monty left me and went directly to the executive officer's office where a large gathering was still trying to figure out what to do about the "man overboard." He received a thorough dressing down by a relieved (and somewhat chastened) executive officer. In describing the tongue-lashing to me later, a twinkle came into Monty's eye and he said, "I'll bet he never announces, 'Liberty, but no boats again.'"

Monty had class! It was with genuine sadness that I learned some years later of his violent death in an airplane accident while testing an F4D Skyray gun system in the skies high over the California desert test range at China Lake.

# PART II
# FLEET TRAINING

*"In any air war, the best trained force always wins."*

*– Colonel Dotan, Israeli Air Force 1971*

*Time and again the Israeli Air Force has proven that, given a reasonably level playing field, the best trained force always wins. Training is that much more important than numbers and quality of arms.*

*When Bob Foley and I walked into his office to interview Colonel "Baban" Dotan in Israel in 1971, he had just returned from a combat mission during which he shot down his fourth MiG (an Iraqi airplane). During the engagement, Colonel Dotan took an ATOLL hit in the tailpipe of his Mirage fighter and the engine flamed out. Unable to relight the engine, the Colonel dove his powerless airplane steeply toward an abandoned airstrip in the Sinai Desert, keeping his speed up so he could continue to evade his pursuer. After successfully "dead sticking" his airplane onto the landing strip, he was picked up and returned to the airfield, Ramat David, in a helicopter. Still sweating profusely, he agreed to an interview with us. We were interrupted several times by his fellow squadron commanders who came by to extend their congratulations and drop off the standard "shootdown" gift, a bottle of fine domestic Israeli brandy. At the end of the interview, he stood and summarized the events of the day with the following announcement, given proudly: "They are not shooting down Baban (nickname meaning Baby) today."*

*By any measure, Baban survived the encounter by superior airmanship . . . the result of extensive and realistic training. Superior training is the most effective arrow in a warrior's quiver!*

*"Baban" was killed later that year in an airplane accident.*

# 6

## "SPAD"

The Skyraider, affectionately known as the "Spad," had become a legend in its own time. The idea for the Skyraider was spawned in the fertile mind of that aircraft design genius, Ed Heinemann, the same mind which later produced the Skyhawk. The year was 1944 and Heinemann, at the time, was the Chief Engineer for the El Segundo Division of the Douglas Aircraft Corporation. The United States Navy was seeking a replacement for the Douglas SB2C dive bomber from two other aircraft companies and had not even invited Douglas to bid.

Approval was grudgingly given by the Bureau of Aeronautics to a Douglas request to compete, but only if preliminary material showed promise and could be produced immediately. The Bureau of Aeronautics official knew that Douglas had not done enough preparatory work and felt certain that they had effectively ruled Douglas out of the competition. They obviously had not figured on Ed Heinemann's genius, boldness and tenacity.

Ed and two other engineers worked through the night and produced preliminary sketches of sufficient quality for the Navy to allow them to enter the competition, but with the stipulation that there be no relaxation of deadlines and milestones. Thus, the aircraft design effort for the XBT2D-1 was kicked off with the Douglas Aircraft Company left in the starting gate and its two competitors off and running, both with an irretrievable headstart.

The first flight of the XBT2D-1 occurred on 18 March 1945, an amazing nine months after that exhausting but prolific night spent by Ed Heinemann and his two associates, Gene Root and Leo Devlin, in a Washington, D.C., hotel room.

The results of the first flight were so promising that a letter of intent to produce 548 of the Dauntless II aircraft was signed 5 May 1945. In April 1946, the

*Douglas A1H Skyraider – "THE SPAD." (Official U.S. Navy photograph)*

U.S. Navy changed the name to the now legendary AD-1 Skyraider. Over the next 12 years of production, the U.S. Navy came up with new missions for this versatile engine-airframe combination. Anti-submarine warfare and airborne early warning, as well as electronic countermeasure missions were added, with modifications growing to four in some versions. In all, a total of 3,180 ADs were built in 7 different models and 28 different series fulfilling a variety of missions beyond the wildest expectations of Heinemann and his two engineers. There seemed no limit to the additional tasking placed upon this able platform. The original maximum gross take-off weight, 18,000 pounds, was increased to an amazing 25,000 pounds. This single-engine attack airplane could carry more bombs than a four-engine World War II B-17 strategic bomber!

Although day attack, all-weather attack, airborne early warning and electronic countermissions were the four basic missions for the various models of the Skyraider, many more were added when it became obvious that the "Able Dog" could "hack it." Anti-submarine warfare, aerial refueling, search and rescue, carrier on-board delivery, photography, rescue combat air patrol and ambulance were also added. But the primary mission of the AD was day attack using bombs, rockets, torpedoes and 20mm cannon. All of the seven basic versions of the Spad, AD-1 through AD-7, were equipped to carry out the day-attack mission. Modifications of these basic models were made for specialized missions.

For example, in 1947 the AD-3N (night attack) and AD-3W (early warning were modified to the AD-3E (search) and AD-3S (submarine attack) to test out the idea of a pair of Skyraiders doing hunter-killer operations against surfaced or snorkeling diesel submarines, day or night. This prototype evaluation concluded that a single multi-place aircraft could do the job better. Thus, the AD-5 was born.

Probably the single most impressive feature of the Skyraider was its power plant. A Pratt & Whitney R-3350 reciprocating engine turned a huge four-bladed 16-foot propeller to deliver 2700 shaft-horsepower at takeoff power. With a wing span of only 50 feet and an empty weight of 10,560 pounds, the AD-1 could achieve a maximum speed of 325 knots, a service ceiling of 26,000 feet and a combat range of 1350 nautical miles. The astonishing thing about the evolution of the AD-1 to the AD-7 was the fact that the empty weight increased by less than 2,000 pounds while the maximum gross take-off weight increased by almost 7,000 pounds.

The Spad was also equipped for special weapons delivery. The AD-4 was modified to deliver nuclear weapons and called the AD-4B. This was about the time of the Korean Conflict, and AD-4Bs were further modified to operate in cold weather. Anti-ice and de-icing modifications were made by then in the AD-4Ls and AD-NLs.

The Skyraider first saw combat in the Korean War where, on 3 July 1950, AD-4s from Attack Squadron FIFTY-FIVE in Air Wing Five, flying from USS VAL-LEY FORGE, attacked airfields and installations around Pyongyang. Attack Squadron ONE HUNDRED FIFTEEN (VA-115) in Air Group ELEVEN, flying AD-4s from the USS PHILIPPINE SEA, lent support to U.S. ground forces during the Pusan action. The USS LEYTE with Air Group THREE, including Attack Squadron THIRTY-FIVE (VA-35) and Air Group TWO with Attack Squadron (VA-65) joined the fray off Korea. In the winter of 1950 USS PRINCETON, with Air Group NINETEEN onboard including Attack Squdron ONE FIVE NINE (VA-159), joined in Korean air action.

When the reserves were called up shortly thereafter USS BOXER and USS BONHOMME RICHARD arrived off Korea with reserve AD squadrons SEVEN HUNDRED TWO (VA-702) from Glenview, Illinois, and NINE TWENTY THREE (VA-923) from St. Louis on board, respectively. AD night attack aircraft detachments from Composite Squadron THIRTY-FIVE (VC-35) supported these air wings. During the Korean War, the Skyraider was acclaimed as the most successful airplane of the war. It was the only U.S. airplane that could carry and deliver accurately the 2,000-pound bomb effectively against interdiction targets such as tunnel entrances, bridge abutments and cave entrances.

When the Skyraider showed up in Southeast Asia, it was in the form of AD-5s modified with dual pilot controls (now designated A-1E) and introduced into the Royal Vietnam Air Force (RVNAF) by the United States Air Force Second Air Division at Saigon in June 1964. These A-1Es were later supplemented by the A-1H (formerly AD-6). Navy Skyraiders were called into combat action in August of 1964. Attack Squadron FIFTY-TWO (VA-52), an element of Carrier Air Group FIVE, flying from USS TICONDEROGA, attacked some PT boat bases in North Vietnam after the attack on the USS MADDOX.

The presence of Skyraiders over North Vietnam increased until the spring of 1965, when an A-1H flown by Commander John Clement Mape, the C.O. of Attack Squadron FIFTY-TWO, an element of Carrier Air Group FIVE flying from USS TICONDEROGA, was shot down by a SA-2 surface-to-air missile (SA-2) over Route Package II in North Vietnam. As the numbers of SA-2 SAM sites increased, the particular vulnerability of the slower, propeller driven A-1s finally became obvious. The decision was made to limit Skyraidera operations over the beach in North Vietnam to Rescue Combat Air Patrol and coastal reconnaissance only. This decision, though obviously a correct one, was greeted with great dismay by Spad drivers the world over. It was the end of a legend.

I had been awestruck by the powerful Skyraider since the day I first saw it at Naval Air Station Corpus Christi, Texas. Early models of the airplane were used in the advanced phase of naval air training for pilots who were programmed into the medium attack pipeline. When I arrived at the Fleet Air Gunnery Unit, Pacific at Naval Auxiliary Air Station El Centro, California, for a tour of duty as a weapons delivery instructor, opportunity knocked. Although I was a jet airplane instructor, I asked for and got permission to "check out" in the Skyraider. I read the pilot's handbook, took a written test, and then Lieutenant Wayne Smith walked me out to the airplane, kicked the tires and then stood on the wing while I sat in the cockpit and discussed operating procedures. An hour later I hit the starter switch and the awesome 16-foot, four bladed propeller started to turn. After completing all the orientation procedures recommended by Wayne Smith, I found myself cruising over the lush Imperial Valley and getting a little bored. For lack of anything better to do I dialed in the UHF radio frequency used by the jet flights on the aerial gunnery range. It was the one week recess between classes and some instructors were out on the gunnery range shooting at a banner. As I listened to them calling "in" and "off" the banner during their shooting runs, I got a brilliant idea. Adding full power to my trusty Spad, I began the long slow climb to 25,000 feet. Spads rarely climbed to altitudes anywhere near 25,000 feet. They were attack aircraft and rarely got above 10,000 feet, and that was only when cruising on long, cross-country trips. I finally reached 25,000 feet and began circling the "chocolate drop," a small black knoll on the desert floor which marked the entry point for flights going into the Chocolate Mountain aerial gunnery range.

In about 15 minutes I heard the flight leader call "off and switches SAFE" as the flight completed the circuit around the range. Looking down I could see the swept wing North American FJ-3 jets cross over the Chocolate Drop and start a gentle turn towards NAAS El Centro. Two other FJs cut across the circle in a running rendezvous while the fourth airplane joined up on the tow airplane to escort it. I rolled the Skyraider inverted, set normal-rated power on the throttle and began a 15,000 foot vertical diving attack on the jets. The white fangs came out!

"These jet pilots have never been bounced by a Spad," I chuckled to myself. "They won't know what to do. If they break into me, they're dead." It was standard operating procedure (SOP) to fly to and from the range in a combat spread formation, simulating a combat situation. All flights were considered "fair game" once they called off the range.

The airspeed indicator on my Spad had read 140 knots as I rolled inverted. Passing 12,000 feet it read 375 knots. The slip stream was howling by the canopy. Directly below me was the number three FJ-3 in the crosshair of my gun sight. This Spad had probably never gone that fast.

"Red One, this is Three, bogie seven o'clock high, break left" came the excited voice from one of the two trailing FJ-3s. The lead FJ rolled into a steep left bank and broke into me. That was just what I wanted as I closed to lethal gun range. I settled into a position 200 feet behind the FJ pulling about five gs, with the gunsight crosshair on the Fury's canopy. I pushed the microphone button in and transmitted, "Red One, you're dead."

When the word got out that Lieutenant Commander "Herb" Hunter, jet fighter tactics instructor and former Blue Angel, had been nailed by a Spad, he never heard the end of it.

Never was the awesome power of the Skyraider so dramatically demonstrated than in the summer of 1958 at the Fleet Air Gunnery Unit at NAAS, El Centro, California. Two young jet instructor pilots, Lieutenant (junior grade) Jim Warnke and Lieutenant Vince Doheny, were scheduled to fly a two-seat Skyraider from El Centro to the Los Angeles International airport to pick up some very important, high-priority airplane parts which had just arrived there by air shipment. The parts were needed in a hurry, so the side-by-side seat version of the Skyraider was employed, as it frequently was, to make the administrative supply run. Jim Warnke, a former Spad driver, was the assigned pilot and Vince Doheny rode the right seat just for the ride.

I remembered my mother's often repeated expression when describing the dangers of assigning tasks to young boys without adequate supervision. "One boy is a boy. Two boys are a half a boy. Three boys are no boy at all." I often thought how apropos that axiom was when applied to young, unsupervised aviators. The flight to Los Angeles International airport and the parcel pick up were uneventful and went as scheduled. The pilot had filed a visual flight rules [VFR] flight plan round-robin to Los Angeles with a 15 minute stop-over for the parcel pick up. Their return route was direct from the airport to NAAS El Centro at an altitude of 10,000 feet. During the return leg it was getting close to sunset and the eastern slopes of the San Gabriel Mountains were cloaked in the late afternoon shadow. Jim Warnke was waxing eloquent to his cohort about how it was "in the good old days flying the Spad" before he transitioned to jets. Jim even offered to demonstrate to Vince how they used to fly low-level practice nuclear bombing missions using the terrain features of mountains to hide from enemy radars. A bored Vince eagerly accepted. Soon they were traversing the ridge line at about 50 feet over the terrain and 220 knots, popping up over rock outcroppings and diving down the other side clinging close to the ground. They popped over one shoulder of a ridge and dumped it over the other side in a steep dive. Jim noticed that the rapid change from bright light to shadow made it hard for his eyes to adjust as quickly as his speed, altitude and good sense might dictate. He had just decided to knock it off when, to his horror, he distinguished directly in front of him a virtual forest of power lines and their supporting towers. It was a feeder line from the Hoover Dam power plant to a substation feeding the southern portion of the Los Angeles basin. Vince saw them at the same time and managed to get out a terror-stricken, "Look out!" There was no time to avoid the wires, and all Jim could think of doing was to advance everything in his left hand (throttle, propeller pitch and fuel mixture) to the firewall just as a huge power line made contact with the spinner on the propeller hub. Vince later described to me the series of sparks, blue lightning bolts and electrical arcs that leaped across the cockpit in all directions and from every angle, literally standing his hair on end as that enormous Pratt and Whitney engine churned its 16 foot, four bladed propeller through the maze of power lines like a hot knife through spaghetti. "It was like a scene from Doctor Frankenstein's laboratory," he said.

By some miracle the battered Spad emerged from the other side of the steel jungle with flying speed . . . but just barely. The plane was a wreck, limping along in a gradual descent at 130 knots. The engine was running at full-throated roar but was making weird sounds because of the terrible damage wrought upon the propeller blades by the power lines. All electrical wiring was fused and all electronic equipment was burned out. The trailing edges of the rudder, flaps and wings were burned, melted and warped. Fortunately, it was a downhill, straight flight path to the small runway at Palm Springs Municipal Airfield, and they barely made that. The long distance, collect phone call that Jim made to the commanding officer of the Fleet Air Gunnery Unit, Pacific started out with, "Hey, Skipper, I've got some bad news for you. . . but Vince and I are okay!"

The conversation went downhill from that point. But the legendary Skyraider had added another page to its already impressive record of achievements. It had snuffed out all the lights in the southern Los Angeles basin with a couple of sweeps of its huge propeller blades, and the aircrew had lived to tell the tale!

# 7

## THE MOON AND STARS

The call for me came into the duty officer's desk at about one o'clock on a Friday afternoon. "It's for you, and she sounds nice," called the duty officer in a voice too loud. He was clearly attempting to embarrass me in front of my peers . . . and he was succeeding. More than just a little nettled, I took the call. Nancy's voice on the other end always made my heart skip a beat or two. This time she sounded excited.

"I'm about to board a Beechcraft headed for North Island. Can you meet me?" she asked.

"Of course." I responded. "I'll be there with bells on." She rang off in a hurry as we both heard the engines of that venerable old SNB Bugsmasher somewhere in the background, coughing into life. Her last words, before she hung up seemed suspended in the air like a faint wisp of perfume in an empty room. She had said, simply, "I love you." God, how wonderful those words sounded to me! As I absentmindedly cradled the telephone I was snapped back to reality by the amused expressions on the faces of my audience, five or six young pilots lounging idly and grinning at me. It was common knowledge that I was smitten, engaged and soon to be married to that cute little number who ran the aviation physiology training unit at Alameda.

It took but a quick calculation of the Bugsmasher's flight time from Alameda to North Island (400 miles), and my driving time from El Centro to North Island (130 miles), to conclude that I must leave immediately by car to beat her to North island. As I was arranging with the duty officer to take the rest of the day off, he asked, innocently enough, "Why don't you take the Beechcraft (we had one also), go pick her up at North island and bring her here? It would take a lot less time." I

*Lockheed T-33 – the U.S. Navy version of the F-80 "Shooting Star."*

hadn't thought of that! What a great idea! Most young pilots only dream of taking their sweethearts for an airplane ride. Hardly any of them ever get to do it . . . unfortunately, or otherwise.

The Fleet Aerial Gunnery Unit owned two Beechcrafts and used them for utility and support missions to transport people and aircraft parts to and from its remote location 130 miles east of San Diego in the Imperial Valley. When I inquired, the schedules officer informed me that one of the Bugsmashers was undergoing a periodic maintenance inspection, but the other was not scheduled for any further flying that weekend. Therefore, it could be made available to me if I could get the commanding officer's permission, and if I could find a co-pilot, at this late hour.

First, I needed to ask the CO. His secretary told me he had already gone home since he was hosting a squadron party at his home that evening, to which I had been invited. I called his home.

"Satterfield's," came the Skipper's voice. I thought I detected the tone in his voice that told me he may have already had his first drink of the evening. He tended to get very witty, informal and good-humored under those conditions. Nevertheless, after listening to my request his response was emphatic.

"Sorry, Paul," he answered rather bluntly, "It's against the rules and I simply can't do it."

"But, Skipper," I pressed. "She's a naval officer, a Lieutenant on active duty. It's perfectly legal for her to fly as a passenger in the Beechcraft on a 'space available' basis. All I need is your permission to go and get her. We can chalk up the mission as a night, cross-country training flight for me and my co-pilot."

"Oh. Okay," he replied, somewhat hesitantly. "Go ahead and tell the training officer that you have my permission. I'll meet her tonight at the party, won't I?," he asked.

"Yes, Sir. You certainly will. Thanks, Skipper." He rang off, and I started looking for a co-pilot. This would normally be a difficult chore this late on a Friday afternoon. All of the "brown baggers" (married pilots) would be planning to attend the Skipper's party. So also would most of the bachelor pilots. The rest would already have their dates lined up.

As it turned out, the task was impossible. Reluctantly, I advised the duty officer that I wouldn't be using the Beechcraft after all. It was already too late to drive to North Island and get there before Nancy landed. I was in a panic, when the duty officer came up with another "solution."

"Well, why not take the T-33?," he asked brightly. (I think he may have been subtly pulling my chain, knowing I was already in a near panic). "You don't need a co-pilot, and it would be one hell of a lot quicker." I thought about that and became extremely doubtful that I could ever get the Skipper to give me a second "permission" . . . especially in a jet airplane like the tandem-seat T-33. However, after giving it some thought, I decided to give it a try. What the hell, I thought. All he can do is say no.

"Satterfield's," the voice said on the other end of the line. This time I was certain that he had already had at least one, perhaps even two drinks. His voice was ever so slightly slurred. Now was the time to strike, I decided and, taking a deep breath, gave it my best shot. So, I asked him.

"Sorry, Paul," came the immediate response . . . although somewhat apologetically. "You know I can't let you pick up your fiance in the T-33 with an ejection seat, oxygen and all that stuff." I decided to throw caution to the winds and press him one more time in hopes that the second martini would kick in and turn him suddenly benevolent.

"Skipper, I don't think you understand," I persisted. "This young lady is a fully certified naval aviation physiologist. She runs the physiology training unit at Alameda and trains all Pacific fleet jet aviators in oxygen procedures, low pressure chamber and ejection seat training." Knowing full well I was treading on thin ice, I pressed on boldly. "She probably knows more about those matters than you do, Skipper." I held my breath waiting for the explosion that was sure to follow such a brazen statement. None came. Instead I heard an answer I could scarcely believe. Thank God, I thought, for the beneficial effects of Bombay Gin!

"Okay, Paul. You can take the T-33. But, come straight back to El Centro. No screwing around. Understand?" I could scarcely believe my ears. I was actually going to fulfill the fantasy of every pilot who ever lived. I was going to take my girl for a joy ride in a jet plane . . . and at night, no less! It was an opportunity beyond my wildest dreams.

The flight to Naval Air Station North Island took only 15 minutes. Since I didn't have to pick up any fuel, I went straight to the flight operations duty officer's desk and learned that the Beechcraft from Alameda was inbound with an arrival time of 15 minutes from then.

When the Bugsmasher taxied into the transient line parking area, I walked out and met Nancy with a big hug and a kiss. I saw the question in her eyes when she saw that I was dressed in a flight suit. I explained that we were going to fly to El Centro rather than drive, and her lack of enthusiasm disconcerted me slightly. When the airman handed a huge suitcase down from the Bugsmasher I knew I was in trouble. There was no place in that tiny T-33 for a suitcase even one tenth the size of that monster. I also knew that I might have some trouble getting Nancy out of North Island without some inquisitive official sticking his nose into the business and causing trouble. In those days, women just didn't fly in Navy jet airplanes!

I had already filed a flight plan listing my passenger as "N. Murtagh, Lieutenant, U.S. Naval Reserve." I took her into the passenger lounge and gave her the flight suit which I had "borrowed" from Red Isaack's locker. In my haste, I hadn't checked it out very well. Red was the shortest pilot in the squadron . . . so I borrowed his flight suit, gloves, helmet and oxygen mask. I don't suppose Red sweated any more or less than the rest of us.

Nonetheless, I noted, the suit smelled fairly rank and I concluded he hadn't washed it in quite a while. While Nancy was changing clothes in the ladies room I ran out to the transient line with her suitcase. I popped it open and began stuffing handfuls of clothing, including shoes and lingerie, into the ammunition cans, on top of the battery and in various parts of the electronics equipment bay. By the time Nancy reappeared from the ladies room, I had stowed the contents of her suitcase in a half-dozen spots in the nose compartment of the T-33 and then hidden the empty suitcase behind a counter in the passenger lounge.

When I told the transient line crew not to come out to my airplane until I signalled with my flashlight, it evoked quizzical expressions of doubt. I didn't want anyone to see that I was sequestering a pretty young woman into the back seat of my airplane. Since women didn't fly in Navy jets in those days, it didn't matter whether or not I was legal. Any scrupulous operations duty officer would stop me. And, since I couldn't prove that it was legal, our little escapade would be stillborn. I couldn't chance discovery at this juncture.

Once I had strapped Nancy into the back seat, put on her helmet and oxygen mask and shown her how to pull up the ejection seat device (imagine!), I signalled with my flashlight for the starting crew. Taking one last look at her, stuffed into the back seat with nothing but a pair of eyes showing, I noticed the dainty white hands with an engagement ring and told her to put on Red Isaack's grimy leather flight gloves. Then I strapped myself into the front seat and waited for the arrival of the sailors who would send us out. Dim silhouettes appeared out of the night and started up the electrical power cart. I gave the signal for electrical power and they plugged us in. A dozen lights came on in the cockpit and everything was ready for a start, except that I couldn't get Nancy to answer up on the intercom. Although I had set up the switches for her in advance, I had failed to tell her she had to depress the button on the throttle to talk to me. By now the noise from the power cart was deafening and I was getting extremely nervous. There was nothing I could do but motion for one of the starting crew to climb up on the wing so I could shout in his ear. He climbed up as requested and I shouted to him to throw the hot microphone switch in the rear cockpit. He crawled back to the rear cockpit and leaned in with his flashlight trained into the cockpit. I held my breath. If only she had the gloves on we had a chance. If not?

A silent prayer crossed my lips. "Oh please, God, I hope she has her gloves on and he doesn't look in her eyes." I couldn't imagine any red-blooded American sailor mistaking those sultry brown eyes and long black lashes for those of a male of the species. Apparently he did not. I heard the hot mike come on and heard Nancy's nervous breathing. Craning my head around I looked to see what the crewman was doing. He apparently saw the hands and then trained the flashlight on her face. Mind you, their faces were only inches apart at the time. I knew that despite the oxygen mask and helmet, one look in those eyes of hers would give away the show. The sailor looked in her eyes and then began shaking his head as he looked inquiringly at me.

Frantically, I gave him the get away signal and the signal to start engines. The sailor, now shaking his head vehemently, climbed down the ladder, removed it, walked to the port main wheel and yanked the line pulling the chocks.

"Can you hear me, Nancy?" I asked on the intercom.

"Yes," came the subdued answer.

"Are your hands clear? I want to lower the canopy and start the engine."

"Yes," again, and it was equally subdued. I started the engine and the shrill whine turned to a roar until the canopy snapped shut closing out the noise and locking us in to a much smaller and more intimate world.

Without further ado, I called North Island ground control for taxi instructions. The response was almost immediate and I stuffed on a handful of throttle, pulling out of the parking ramp with a roar of power and a cloud of dust. Ground control cleared me to taxi to the takeoff end of Runway 29, the western runway . . . the one pointing directly at Point Loma. The wind was calm, they informed me, as they told me to shift to tower frequency for takeoff clearance. We taxied out with a roar of power and headed for the duty runway. Maybe I was going to pull this off after all.

North Island tower cleared us on to the runway and cleared us for immediate takeoff. I rolled out onto the runway and came to a halt for one last check of the takeoff checkoff list.

"Are you ready to go flying?" I asked my passenger over the intercom.

Another subdued "yes," was my answer as I rammed the throttle to the firewall and released the brakes. The three of us, my trusty T-33, my future wife and I, started slowly trundling down the black runway toward the dark mass that made up the heights of Point Loma directly in our path. It was a hot and humid night. This combination meant that our underpowered jet would eat up most of the 7,000-foot runway and very clumsily claw its way into the damp night air.

A prudent pilot, under these circumstances, would leave his flaps down after takeoff, make the required 90-degree left turn to avoid flying over Point Loma and then raise them when he had a comfortable margin of airspeed. But that was a "pussy" maneuver. A "real fighter pilot" always sucked up his flaps immediately, accelerated to climb speed and turned to his departure course. Of course, that would be stupid under these circumstances. But, this was a dream flight! It had to be flown with panache. After all, it was my first, and probably my last, opportunity to impress my future wife with my skill as a jet pilot extraordinaire!

Shortly after the T-33 staggered into the air, I yanked up the flaps. The poor T-33, with barely enough flying speed, settled sickeningly down toward the black channel water while the monstrous dark bulk of Point Loma rushed at us at 150 miles per hour. The course rules for departing flights prohibits flying over Point Loma. But, I now had no choice. There simply was not enough flying speed for me to undertake the required left 90-degree turn. I held my breath, praying for the airspeed indicator needle to start increasing toward the 200 knot mark. No way! Too late! I began an excruciatingly gentle climbing turn to the left, almost brushing the tops of the trees on the eastern slope of the point.

Nancy sensed we were in trouble, I think. I could tell from the increase in her respiration rate . . . her breathing over the intercom. We made it with only about 50 feet between us and the housetops along the crest of Point Loma. Her breathing only began to subside after the lights of Cabrillo Lighthouse flashed just beneath us and we were free over the ocean. The remainder of the departure was uneventful.

"Butch" Satterfield's admonition to proceed direct to El Centro "without any screwing around" vanished from my consciousness like a wraith as our sleek jet, now with plenty of airspeed, climbed like a homesick angel into the brilliantly star-studded sky. God, it was beautiful! How romantic can it get?, I wondered, as we climbed through 20,000 feet and headed north toward the Los Angeles basin. There were ten million lights in the depression of that megalopolis. Over head were another billion stars. We winged our way over Los Angeles, the engine now a faint whisper and the stars drifting slowly past our canopy, and we turned east toward Palm Springs. I talked like a tour guide, pointing out the various scenic points of interest, but Nancy was strangely silent. I made the stupid assumption that she was so overcome by the beauty of the scene that she was speechless. Finally, I lapsed into an appreciative silence as we turned south over Palm Springs toward the Mexican border at Mexicali.

We cruised south, the engine a faint whisper. The sparse lights of Imperial valley were a dark carpet drifting slowly beneath us. Above us, with no apparent motion at all, the night sky crept. Each star seemed as big as an apple. I felt as though I could have reached out and grabbed any one of them. It was exquisite . . . millions of diamonds scattered on a black velvet backdrop by some profligate god! I was in seventh heaven as we approached the Mexican border and turned back westward toward El Centro.

We began a steep descent about 30 miles east of the airfield. Realizing that my flight of a lifetime was about to end, I decided to give it a memorable, final flourish. El Centro tower cleared us to land and we entered the "break" at 400 knots, wrapped it up into an 80-degree bank, pulled six gs turning to final and glued it on the runway at 140 knots like a pro! Now that, I thought, will be a flight for both of us to remember!

As the engine wound down I climbed out of the front cockpit and went back to put the safety pins in Nancy's ejection seat. I then helped her climb down to terra firma. My line chief petty officer greeted her royally and I whisked her off to her motel where she changed for the party. Of the drive to the Satterfield's house, I have, strangely, no recollection. Perhaps it was anti-climax. On our arrival at the party, Nancy was the subject of great interest. After all, there was not a woman at the party who could boast of having done what Nancy had just done.

It is strange, in hindsight, that I cannot recall any details of that party, other than that Nancy received a great deal of scrutiny and attention. The biggest surprise came years later when my wife, in a moment of extreme candor, admitted to being terrified throughout the entire flight. She had her eyes tightly shut, she told me, throughout the entire one-hour grand tour of Southern California. I thank God, and her, that she waited a decent number of years before bursting my bubble! If she hadn't waited so long, I'm sure my fighter pilot's ego would never have survived such a devastating blow. She must have instinctively known that. How frail we mere male mortals be!

# 8

## THE PASSENGER

The note from the Skipper's office was both peremptory and urgent. I hurried over as soon as I saw it on my desk. I had just returned from a gunnery flight debrief and saw, from the time on the note, that it was already an hour old.

"Paul," the C.O. said, "I want you to take Captain Vredenburg to Washington. He needs to get there in a hurry. So you'll leave this afternoon as soon as you can get ready."

"Who is Captain Vredenburg?" I asked. The C.O. made a wry face and responded caustically.

"He's the commanding officer of the air station." The Skipper went on to explain that Captain Vredenburg had to report to Washington as a member of a selection board and couldn't get there in time via the airlines.

I was further told that my passenger had never flown in a jet airplane and therefore had received no formal aviation physiology training. I was cautioned to be on the alert for any problems which this ignorance might present. I was enjoined to be extra detailed in my pre-flight briefing to include complete instructions on how the oxygen and pressurization systems worked.

As I walked out of Commander "Butch" Satterfield's office, I was beginning to regret having been selected for this "honor." Although it represented an opportunity to get some extra flight time, it was, I knew, boring flight time. Furthermore, my wife wouldn't relish the thought of spending several evenings home alone with the two infants, especially on such short notice.

I called her, told her about the trip, and asked her to pack my overnight bag. My prediction was accurate. She didn't like the idea. But she accepted it and packed my stuff. I picked my bag up an hour later after I had completed a rather hasty flight planning drill and filed a flight plan.

I could read it in his eyes the moment we met. He was asking himself why my Skipper had entrusted the safety of his body to a young Lieutenant (junior grade). The expression on his face annoyed me, but I tried not to let it show as I sized up my passenger. He was short, heavy set, a little fat with a deeply creased face, clear blue eyes and salt-and-pepper hair. I judged him to be in his mid-50s (old, by my standards then).

A last minute, one day delay in the convening date of the selection board caused us to leave at dawn on the following day. The departure from NAAS El Centro was uneventful. Weather was forecast to be good all the way east to the Allegheny Mountains. From there on we could expect to encounter increasingly foul, rainy weather all the rest of the way to Andrews Air Force Base, on the outskirts of Washington, D.C.

When we leveled off at our cruising altitude of 43,000 feet, we seemed to be standing still. There was no sensation of motion . . . like hanging in space. Far below, the drab sameness of the great American desert seemed to inch by almost imperceptibly. My passenger was not talkative. For that I was grateful because as a general rule I don't like garrulous traveling companions.

Prior to takeoff, I had set my rear-view mirror so that I could see his eyes above the rubber oxygen mask. With the sun visor of my helmet lowered, I could watch his expression without him seeing mine. His left hand was draped rather casually over the canopy rail and I assumed (nay, hoped) the other was holding the control stick since I had just given him control of the airplane. I thought he might enjoy it.

Surely, it was a coincidence that I handed the controls over to him just as the fuel level in the tip tanks had dropped to just a little less than half. This is the one condition which makes it almost impossible to trim the airplane for level flight . . . and go hands off the controls.

There were no swash plates (baffles) in the tip tanks. Therefore, when they were half full, the fuel in them tended to slosh back and forth, causing a slight, and almost gentle rocking of the wings. No amount of lateral trim would stop it. The only solution was to apply ever-so-slight lateral stick corrections and continue to fight it until the level of fuel in the tanks dropped a little lower.

If a pilot really wanted to take all the motion out of this phenomenon, he could do it best with both hands on the stick. Most of us thought it wasn't worth the candle and did a reasonable job one handed . . . but it took a little practice. I was just annoyed enough at my aloof passenger to put him to the test.

"Would you like to fly it, Sir?," I had asked him over the intercom. He condescended and responded to my gentle stick shake with one of his own. I noted with great amusement that his left hand remained casually draped on the canopy rail.

The rocking began and I switched off the "hot mike" so that if I laughed he wouldn't hear me. Gradually the magnitude of the rocking excursions increased. I could see that he tried to correct the problem with the lateral trim button on the stick. When that failed, he tried to control the rocking with stick inputs.

I rested my own hand on the stick, ever so gently, to derive the maximum amount of pleasure from his discomfiture. I could feel his inputs, always a little too much and always just a shade out of timing. The rocking increased to a point that he forgot all about holding altitude and we began to climb and descend in a rocking, wallowing motion like a drunken pelican. I was by now laughing so hard that I was afraid my passenger might notice my shoulders shaking from the rear cockpit.

But never did he move his left hand from full view on the canopy rail. However, I noted that the left hand was now gripping the rail tightly; and the knuckles in the tight fist were growing white. He never said a word. I couldn't have responded calmly if he had.

Finally, I asked him if he would like to give control of the airplane back to me. This time there was no condescension in the turn-over. But I could feel his frustration, embarrassment and irritation. It hung in the air, unspoken, but so heavy one could have cut it with a knife. I began our descent into El Paso, Texas, with the weather so clear we could see the airport from 70 miles away at cruising altitude.

"Now don't run us into a mountain," he admonished testily as I thumbed out the speed brakes and began a rather steep descent. I promised to be careful. The refueling on the ground went quickly and smoothly. I oversaw the operation while Captain Vredenburg went to the transient lounge to make a few telephone calls, much in the manner of someone directing the affairs of state.

Our departure from El Paso was also uneventful, with our next destination Naval Air Station Memphis, Tennessee. The weather was still fairly good. This time, when I leveled off at cruising altitude and offered him the controls, he declined, gruffly stating that he had to review some notes preparatory for the selection board.

Somewhere during the leg to Memphis, our T-33 developed some minor mechanical problem. I cannot now remember for the life of me what it was. But, I do recall that it was sufficiently serious that I didn't want to attempt an approach into the Washington air traffic control area, and worsening weather, without having it fixed.

When I announced to my passenger that there would be a delay to fix the problem, he directed me to call ahead to Memphis and request an alternate mode of transportation. I did so, and when we taxied into the transient line at Memphis I saw a cluster of minor dignitaries waiting to escort Captain Vredenburg to a waiting SNB, its propellers idling.

It took all of my self control to keep from chortling out loud. The Beechcraft cruised at about 135 knots. I knew it would take him forever to fly to Washington in that ancient crate. He might just as well have taken a train. My passenger climbed out and walked away, never saying a word of thanks or good-bye. I last saw him climbing into the door of the Beechcraft. A mechanic on the transient line fixed my airplane in about an hour. I now felt as free as a bird.

Since the good Captain Vredenburg hadn't given me explicit instructions to return direct to El Centro, I felt a certain amount of latitude as to how and when I got home. I called my brother, Jim, a former Marine Corps officer who now lived in Oak Ridge, Tennessee. Not far from where my brother lived was McGhee-Tyson Air Force Base. I told my brother to meet me at the Air National Guard transient line in one of his old khaki Marine Corps uniforms. I had Captain Vredenburg's flight helmet, oxygen mask and flight suit strapped in the rear seat. I told my brother he was about to get his first ride in a jet airplane . . . albeit an illegal one.

I knew I could pull it off! But, I also knew that when I got home, I would have to explain why I landed at McGhee-Tyson (half an hour further east); and then flew a mission from that base. The original mechanical problem, I decided, would have to recur, requiring me to land to get it fixed . . . and then a special test flight from McGhee-Tyson to ensure the fix was working okay, before returning to El Centro. I decided to not discuss the relative locations of McGhee-Tyson and Memphis in hopes that the people at El Centro would not notice anything unusual about my movements on that particular day. No one ever caught on.

My brother and I rendezvoused as planned and I had the great pleasure of showing him what it was like flying in a jet. We went sight-seeing, did some simple acrobatics and returned an hour later having done something I had always wanted to do.

In 1958 there were few if any U.S. airlines equipped with jet-powered airliners. The British had developed a twin-engine airplane, but their acceptance was slow in coming in the United States. In many circles, jets were still considered a little unreliable.

So my brother's jet ride was really quite an adventure for both of us. He had been in the Navy flight training program for a while, but it had been just after the end of World War II and the decision by the Navy to virtually shut down the flight training pipeline caught my brother and many other talented young men in the bind. They were sent home.

Of course, I had to get headed back. I avoided reporting my return plans to my boss by telephone until I had returned to El Paso. Once there, I called him and told him most, but not all, of what I had been doing with his airplane.

By then it was late. I had flown five flights. He told me to remain over night in

El Paso and return the following day. I did so with a reasonably easy conscience. So far, I had gotten away with murder. Tomorrow would turn out to be quite another day!

I spotted him the moment I stepped into the transient operations building at El Paso International Airport the following morning. I had a peculiar feeling that he meant trouble. There were only a few pilots preparing for departure at that time of day, and I sensed immediately that he was looking for me. he was dressed in a disheveled green Army officer's uniform. I could see by its condition that he had been traveling in it for a long distance. It badly needed pressing. I came out of the flight planning room and headed for the weather briefing telephone in the corner of the main room. He followed me and waited while I talked with the weather forecaster about upper winds and weather for my flight to El Centro. The headwinds were not too bad and I concluded that I could make the 850 miles in no more than two and a half hours. There would be adequate fuel reserve and the weather at destination was the usual clear and 75 miles visibility. I sensed the man would address me as soon as I hung up the telephone. He did.

"Sir, are you Mister Gillcrist?" he asked. I nodded and he continued. "I'm on my way home from a tour of duty in Germany. My wife is waiting for me in San Francisco, and I need a ride as far as El Centro if you will take me along." His voice had a plaintive quality and I knew right away that I was going to have a hard time turning him down. I also knew that I should turn him down. There was a sense of accomplishment at having gotten my brother a ride in a jet. It was accompanied by the conviction that I should not push my luck . . . especially not for someone I didn't know, much less care a rap about.

"Sorry, Captain," I responded. "It is against the rules to take a passenger in a jet plane when he hasn't had any aviation physiology training. You could get into all sorts of trouble with your oxygen system and there would be no way I could help you. I hope you understand that you are asking me to break the rules, and I can't do that. Besides," I added without thinking, "you have no equipment."

He responded immediately that he had looked into the rear cockpit of my airplane and seen the helmet, oxygen mask and flight suit securely tied down on the ejection seat. I mentally kicked myself for overlooking such a simple thing. It also made me feel just a little hypocritical since I had just broken those rules in my brother's case.

This put me on the defensive, and I bridled at the pressure he was applying. He pleaded with me and told me how long it had been since he had seen his wife. There was an infant daughter he had never seen, and another son whom he missed dearly. Wouldn't I please reconsider? He promised to listen carefully to all instructions I would give him and make no mistakes.

Finally, I relented and, marching him out to the airplane, gave him a quick but reasonably thorough lecture on the oxygen system, ejection seat and internal communications. I knew I was doing a dumb thing. To counter the guilt, I consoled myself with the idea that I was doing some great corporal work of mercy. I had been in my passenger's position before and knew how disappointing it could be to be told no for no apparent reason.

Underlying it all, however, was the sick feeling that soon I would regret my softness. I was also uneasy about our departure weather. There was a powerful squall line moving toward the field from the northwest, and it was kicking up a substantial dust storm as it approached.

We manned up our airplane hurriedly and, as soon as I could get the engine started, I put the generator on the line, turned on the radios and called ground control for taxi clearance. They advised me that the field was about to be closed due to high winds associated with the approaching sand storm. We looked to the northwest and could see that enormous black, churning cloud approaching. I estimated it was no farther than 10 miles away.

Quickly copying down my flight clearance, I taxied out at best speed to the duty runway, all the time keeping a weather eye on the storm cloud. It looked scary. Fortunately, we were about to take off on Runway Four to the northeast and would quickly be clear of it.

"Sir," came the plaintive voice of my passenger on the intercom. "I just remembered that I have a large, $75 bottle of perfume in the bag which we stowed in the left ammunition compartment. It's my wife's favorite brand, and I'm afraid it might rupture up there in that unpressurized area. Can I please get it out and hold it in my lap for the flight?"

Jesus, I thought, that son of a bitch sure picked the wrong time to start paying attention to my lecture on high altitude pressurization. Damn him, anyway! I just sat there for about ten seconds . . . just glowering.

"Captain," I rasped harshly, "This is a hell of a time to remember the Goddamned perfume! I can't go out there and get it for you because I have to stand on the brakes."

"Please, Sir," he pressed. "I saw how you opened the ammo bay door and also how you locked it. I can go out there, take the perfume out of my bag, lock it up, and be back here in a couple of minutes, all ready to go." His voice fairly whined with pleading. What a wimp, I thought!

"All right," I answered. "Do it. But be extremely careful about closing and locking that door. And, hurry up. There's a humongous sandstorm headed our way and it will hit us in a few minutes. If it does while we are still sitting on the ground, our flight is canceled and I'm going to be one highly ticked off son of a bitch. Do you understand?"

He said that he understood and, with an alacrity which defied his obesity, unstrapped and jumped down to the tarmac. Quickly, he began unbuttoning the large ammunition bay door on the left side of the fuselage just forward of the cockpit. I loosened my shoulder harness and leaned as far forward and to the left as I could while still keeping my feet on the brake pedals. I was able to get a pretty good view of him as he withdrew a parcel from his hand bag, restowed it in the compartment and then secured the door latches. They looked secure to me as he scrambled back into the rear cockpit and began the elaborate process of strapping himself in. Sand was starting to pelt the exposed side of my face as I hooked my oxygen mask back on and lowered the canopy. Powerful gusts of wind had begun to buffet our frail airplane and the forward fuselage began to swing back and forth, the nose wheel castoring gradually away from the direction of the gusts.

"Navy Jet Two Zero One, this is El Paso Tower. You are cleared into position for immediate takeoff. If you cannot do so, hold your position. The field will close shortly due to heavy winds. Over." My response was immediate.

"El Paso Tower, this is Navy Jet Two Zero One. Taxiing into position for immediate takeoff." At the same time, I rasped to my fat passenger on the intercom, "Tell me when you are all hooked up. We have to get out of Dodge right away."

I lined the nose of the T-33 on the white centerline and, holding the brakes, ran the throttle to full power. Again I asked my fat passenger if he was all hooked up and still there was no answer. I was quickly getting exasperated when he finally spoke up.

"I'm all strapped in, Sir." That was what I was waiting for, expecting at any moment to be told the field was closed. I released the brakes and we began rolling down the runway. The radio call from the tower startled me.

"Navy Jet Two Zero One, this is El Paso Tower. Abort your takeoff. The field is closed. Acknowledge. Over." I clearly heard what the man said, but had the presence of mind to respond as the T-33 began to pick up speed.

"Wilco, El Paso Tower. This is Navy Jet Two Zero One. Your transmission is breaking up. Understand cleared for takeoff. Out." I realize, in reading this, that my response may seem a bit cavalier. However, we were almost at the point of no abort. The issue in a few seconds would be moot . . . and besides, my mindset was now irrevocably in the flying mode. My mind shut out the tower's response which was a repetition of the earlier directive. I had rotated the nose of the buffeting T-33 to the flying attitude and was concentrating on coaxing the recalcitrant airplane into the air. The airplane broke ground and I immediately raised the wheel handle. I saw, to my absolute horror, the left ammunition bay door start to open!

I was a very large door, hinged at the top. Fortunately, I pulled my hand away from the flap handle, which I was about to raise, and watched terrified as the door slowly opened wider and higher in my field of view. The airplane began to shudder and started an uncontrollable roll to the right. We were perilously close to the ground; and the airplane had stopped accelerating. When the bank angle reached 45 degrees, the nose of the airplane began to slowly drop in spite of the fact that I now had the control stick almost all the way back in my lap and fully left. I felt utterly hopeless and thought, "So this is what it's like to die . . . and all over a lousy bottle of perfume . . . Jesus Christ!"

Then something came over me. Never give up, I reminded myself as I kicked full left rudder. Don't ask me why I did that. It made absolutely no good sense . . . but I did it! At least I can die trying!

"What's happening?" came the plaintive whine from the rear cockpit. He had repeated the question when I shouted back in outrage.

"Shut up!"

The right wingtip was about to make contact with the ground when I kicked the left rudder. To my amazement the ammo door slammed shut with a loud bang and the airplane rolled wings level.

The tops of the sage brush were drifting past the canopy. I felt like I was looking up at them. We couldn't have been more than 20 feet over the desert floor. I was ignoring a series of frantic radio calls from El Paso tower on the emergency frequency to pull up. I took out the left rudder and, to my horror, the ammo bay door began to open again and the terrible airframe shuddering began again, along with the uncontrollable right rolling motion. I stomped the left rudder again, and again the door slammed with a bang. I found that by holding about half-left rudder throw, and about a 15 degree right bank angle, I could keep the door shut. This, of course committed us to a right hand turn and I recognized the approaching buildings of downtown Juarez. There was a gentle downward sloping of the terrain to the south and the airplane was now about 200 feet above the terrain; and holding about 150 knots.

My mouth was dry. I found I had been holding my breath involuntarily, and let it out with a great sigh. Certain death was no longer imminent!

"El Paso Tower, this is Navy Jet Two Zero One. I have a controllability problem and am declaring an emergency. I have to land immediately. Over."

"Negative, Two Zero One. The field is closed. Out," came the immediate reply.

I was in no mood for a discussion. It was crystal clear in my mind that I was going to land this airplane as soon as I could, regardless of any directions to the contrary. I didn't really give a damn at this juncture. When I conveyed this thought to the tower, they advised me that the field was closed and that if I tried to land, I would be doing so at my own risk! A trivial thought flashed across my consciousness for just an instant. It almost brought a smile to my face. Here I was, a carrier pilot, being told by an Air Force tower operator that the landing I was contemplating must be made at my own risk.

Wasn't that always the case? At whose risk was I usually landing, if not my own? What a dumb thing to tell me! I acknowledged "risk" and continued what was now an exaggerated cross-wind leg to Runway Four. As I rolled out on final (and stopped my continuous right turn) the damnable door would start to open. I would then feed in more left rudder and counter it with right aileron; and the door would shut. Finally I settled on a crabbing approach which kept the door opening and then shutting immediately as I bobbled and swiveled my way down the glide slope.

Up to that point our unguided tour had provided us with a dusty and rain-pelted tour of downtown Juarez. Now, we were back stateside and aimed roughly at the landing end of Runway Four. When I asked for winds, I immediately regretted having done so. After all, I was going to land on this approach regardless of winds or anything else. Winds were variable, they told me, at 40 knots, gusting to 60. The squall line had just passed and the wind direction at the moment was in radical change.

The pilot's handbook tells you never to attempt to land with full tip tanks. Mine couldn't have been fuller. That was a mere detail . . . a bagatelle! Down came the landing gear and I held my breath to see what this squirrelly airplane would do next. The surprise was pleasant. Now, I found, if I held full left rudder and enough right aileron to keep the right wing down about 10 degrees, I could stay on glide slope at 140 knots and the ammunition bay door opening about six inches then slamming shut immediately. This seemed like a delightful compromise to some one who moments earlier had looked death in the eye.

I made the mistake of informing the tower that my wheels were down. This forced them to repeat the litany about the field being closed and my landing would be at my own risk. It sounded to me as though they were more concerned about covering their asses in case we turned into a fireball in the next few seconds than they were about seeing us return in one piece.

As we continued our approach in a crabbing, wallowing motion, violent winds buffeted us like a cork in a raging sea. The violence was beyond anything I had ever experienced before. We bounced up and down, then were slammed sideways and back. I was able to keep in the general vicinity of the approach cone more by luck than any other single factor. My determination to plant this airplane on the runway was also fairly intense. As we touched down, the heavy wingtips flexed downward then upward flinging our little T-33 a good 20 feet into the air. Down we came again, and the cycle repeated itself, this time with less violence. On the third bounce I knew we had it made. Then a violent gust picked up the left wing and our airplane headed toward the left shoulder of the runway despite full right rudder and left aileron. Just as we were about to leave the runway, the gust abated and my control inputs brought the airplane back to centerline as the left main mount slammed onto the pavement. We made it! Several more gusts hit but as we slowed down their effects were easier to control. I turned off the far end of the runway and told the tower that I needed to stay in the warm-up area for a minute and would be off the radio. I didn't even wait for them to acknowledge.

In a rage I opened the canopy, ignoring the stinging blast of sand granules on my face, and told my passenger to get out and lock those God-damned doors. This time I unlocked my shoulder harness and, standing half way up in my seat with my toes still holding the brakes, watched my passenger dutifully lock all four door latches after I assured myself that they had suffered no particular damage during our excursion over downtown Juarez. Finally, the Army Captain climbed back in and I closed the canopy. I could feel grit all over the inside of my flight suit. I cannot recall ever having felt such a powerful combination of misery and rage!

On the plus side, I was alive! . . . and that was quite a plus after all. I keyed the microphone.

"El Paso Tower, this is Navy Jet Two Zero One. I appear to have solved my controllability difficulties and am ready for takeoff. Over."

What chutzpah! I imagined the tower operators were probably laughing their heads off. On the other hand, they were probably in a very grateful mood that we weren't splattered all over the runway in a large ball of fire. Considering all that had happened, I decided that I would be better off as far away from El Paso International Airport as I could get . . . as fast as could get there. I somehow suspect the tower crew felt the same because they suddenly became more accommodating.

"Negative, Two Zero One. The field is still closed. However, the surface winds are abating. You are cleared to taxi down the duty runway to the far end, taxi clear and hold your position. Over."

I was very ingratiating about acknowledging their directions . . . almost unctuous in my manner. We did as directed. It took several minutes to taxi down the long 12,000 foot runway. While we waited, I recalculated my remaining fuel and decided we still had enough for the flight. I also gave a careful study of the appearance of the wings, flaps, tip tanks and ailerons. Everything seemed to be working. Although I had no idea what stress the airplane had been subjected to with the ammo door slam-banging around, there was no evidence that I could see of buckling. We would have the metalsmiths give the airplane a thorough inspection for structural damage when we got back to El Centro.

After about five minutes, the tower cleared us to depart. I'm sure they were glad to get rid of those Navy clowns. The flight back to El Centro was uneventful and relatively peaceful. However, my anger didn't subside so rapidly. When we landed at El Centro, I found myself wanting to get rid of my obese passenger as fast as I could.

For that reason, I asked the tower for permission to drop my passenger at the tower building which also housed a rudimentary air passenger terminal. I waited with engine running while he unloaded his baggage and tossed the helmet, flight suit and oxygen mask into the back seat. He tried to thank me, but I was having absolutely no more truck with this guy. I waved him away rather rudely and added a batch of power when I taxied away. He stood there in a small cloud of engine exhaust and dust looking even more forlorn than ever. His wrinkled green Army uniform looked as though he'd been to the moon and back in it! As I pulled into the parking ramp at FAGU, I said a silent prayer that I wasn't all wrapped up in a pile of wreckage back in west Texas! I've never discussed the incident until now!

# 9

## A VIRGIN BANNER

Certainly one of the most difficult of a fighter pilot's missions is aerial gunnery. It demands the most exquisite degree of discipline, the highest judgment, perfect physical coordination and extremely fine reactions to hit the target. Training in aerial gunnery is accomplished by shooting at a towed target. In 1955, the targets towed at El Centro were made of polyurethane and were 6 feet high by 30 feet long, and were towed at the end of a steel cable 1,000 feet long. Today's banners are still made of the same material, except that they have woven into the fabric a metallic fiber which causes the banner to reflect radar energy. This means that the banner becomes a radar reflector and allows the airplane's fire control system to get a radar lock-on to provide range and range-rate information for a correct firing solution. Furthermore, today's banners are 8 feet high and 40 feet long.

This must seem huge to the casual reader. However, when viewed from the cockpit at a distance of three or four miles (where gunnery runs normally begin), it is much smaller looking than a postage stamp. And, when opening fire at a range of 1,500 feet, they are still not very large to the shooter's eye.

Naturally, the larger banners of today use a longer steel tow cable of 2,000 foot. There is a three foot bull's eye painted in the center of the target. It is so small as to be barely discernable when a shooter commences firing.

As a general rule, the higher and faster a banner is towed, the harder it is to hit. Consequently, we started our students out shooting at banners towed at 200 knots and an altitude of 15,000 feet. As they began to learn how to fly the complicated gunnery pattern, the speeds and altitudes were increased incrementally. The next higher altitude was, for example, 20,000 feet and 220 knots. The most demanding of all aerial gunnery training was to shoot at a banner towed at 25,000 feet and a speed of 250 knots.

In those days when there were no afterburners, and the airplanes were extremely power limited, the margin for error in a 25,000-foot tow was almost zero. Each gunnery run had to be perfect. It was very difficult to hit well under those circumstances, and was the greatest test of aeronautical skill I know of, even today!

A typical gunnery flight consisted of four airplanes and a tow. The pattern they flew looked like a moving figure eight when viewed from above. The gunnery pass would begin at the "perch," a position abeam the tow about one mile and 5,000 feet above it. A nose-down turn would be made toward the tow plane. When the nose of the shooter was roughly pointed at the banner, the turn would be reversed on a downhill pursuit path with the nose of the airplane pointed at the tow from there on. At full power, the Cougar could build up to 450 knots by the time the banner came into firing range (1,500 feet). Closing velocity was high, and minimum firing (break-away) range was reached so quickly that there was time for no more than a two-second firing burst. At minimum firing range, there was just time for the shooter to roll his wings level and pop over the top of the banner.

For safety's sake, the shooter then flew parallel to the banner cable until he passed the tow plane. From that point, he would call "off" and commence a nose-high climbing turn back to the "perch." The whole pattern took about one minute from "perch" to "perch." This meant that a well-disciplined gunnery flight would have a shooter firing every 15 seconds with the four airplanes evenly dispersed around the pattern. It is a thing of beauty to watch a skillfully flown gunnery flight. There are elements of Swiss watch precision, grace, power and speed which are difficult for the average groundling to appreciate.

Although the flight leader had the ultimate responsibility for the safety of the flight; the tow pilot was also a very important safety observer. In addition to keeping his banner, and therefore the flight, inside the gunnery range, the tow pilot also was responsible for seeing that the shooters didn't do anything unsafe . . . like, for example, shooting the tow plane down . . . that has happened.

In a typical gunnery pass, only two radio calls are required of the shooter. When he turns in off the "perch" he calls "in." When he passes the tow, he calls "off." For safety sake there is a minimum firing cone of 15 degrees from the banner. Once inside that cone, the shooter is prohibited from firing. The only two other restrictions have to do with tow plane safety. In the first, the firing airplane must always be descending during the firing run. Secondly, once he reaches minimum firing range (600 feet), the shooter must stop firing and initiate break-away procedures. This has the added benefit of preventing the firing airplane from colliding with the tow. That has happened too!

*Flight of FAGU F9F-8Ts over the Imperial Valley, March 1957. (Author's collection)*

The leading edge of the banner contains a steel tow bar two inches in diameter. Many a gunnery plane has returned from a flight with a tow bar buried deep into the leading edge of the shooting airplane's wing . . . very embarrassing, when not fatal!

After months of continual gunnery flights, the average instructor achieved a level of competence rarely achieved by the average fleet pilot. That is not snobbery . . . it is a fact of life! It is because of this fact that certain unwritten rules existed at the gunnery school. One rule said that the instructor always assigned himself the worst shooting airplane in the flight. Let's face it, certain airplanes routinely "hit" better than others. It quickly became common knowledge at the school. Students would vie for the good shooters. By the same token, certain airplanes were "known" to shoot high or low ahead or behind. Instructors were usually loath to admit it; because the same periodic "harmonization" techniques were applied to all airplanes. But, it was a fact, nonetheless just as certain rifles used by a shooting team are known to score higher than others.

Another rule was that there was always a standing bet by the students against the instructors. If any student got more hits than his instructor, the instructor had to buy the student a round of beer at the club that afternoon. A third rule made it mandatory for any instructor to buy a "round" for the whole flight if he ever came back with "a virgin banner" (with no hits by him).

The banner was held in the vertical plane by the tow bar at the leading edge with a weight at the bottom. As mentioned earlier, a banner collision was considered extremely bad form as well as terribly hazardous to one's health. There was a sign in the squadron ready room which was updated daily. It said: "There has not been a banner collision in this organization since (the date was several years and several hundred thousand gunnery runs ago)." It was an intimidating sign . . . and an effective one, at that!

It was generally understood that the closer a shooter "bored in" to the banner before opening fire, the better chance he had of getting hits. So, for the most part, aggressive pilots tended to ignore the minimum-range rule. With arrival of more reliable gun cameras in the late 1950s, the instructor could review a shooter's gun film and usually nip that dangerous tendency in the bud. Oftentimes, while reviewing gun-camera film, I would catch a shooter opening fire at 1,000 feet instead of the recommended 1,500 feet and breaking off at 300 feet instead of the required 500 feet. This was nothing more than playing chicken at 500 miles an hour. Sooner or later it would catch up to the miscreant . . . and often kill him.

But all these safety rules notwithstanding, I tried my level best never to bring back a virgin banner. That would be a singularly galling experience, regardless of the mitigating circumstances (like jammed guns). One particular day, it happened to me in spite of what I can only describe as a monumental effort to get at least one hole in the banner.

It was a typical hot July day and my three students were about halfway through the five-week syllabus. They were beginning to get hits on a consistent basis. It was about this point that all instructors get really serious about their own shooting efforts . . . for obvious reasons. We were shooting at 20,000 feet and the student's gunnery passes were beginning to look pretty smooth.

I called "in" on the first "live" firing pass and bored in close. There was a good reason for doing this on the first pass. I had learned the trick from another instructor; it was one of those things one never passes on to students! That was because it was dangerous . . . but it worked like a charm, if done properly!

It would be nice to know whether one's 20mm guns were shooting exactly where the gunsight pipper was placed on the target. If in fact they were hitting a little high or low, or ahead or behind, it would be a tremendous advantage to know. My instructor friend explained the trick to me in a strangely matter-of-fact manner.

On the first firing run, all of the holes had to belong to the instructor. After that, there would be no way of knowing even if one could see the holes as he passed the banner. Therefore, all one had to do was "see" the holes after the first run to determine where his guns were shooting. (This assumed, of course, that the shooter had made a perfect run). When I told my friend that I never saw holes anyway, he looked at me in a patronizing sort of way.

"You have to 'pan,'" he told me.

"What do you mean, pan?" I asked.

"Pan, as in a camera," was the reply. He explained that he recovered in close, and flew by the banner close (50 feet away as he went by). Then, by turning one's head fast as the banner went by, one could make the banner seem to stand still for a brief moment. It was a film technique that everyone knew about, he told me as though speaking to a cretin. The foreground is clear as a bell and the background is blurred. The foreground, in this instance, was the banner.

"If there are any holes," he concluded, "they are yours. And, if you made a perfect run and notice the holes high or low, or ahead or in back, then that is where your airplane is shooting. After that, all you have to do is compensate for the error and you never buy beers."

I tried it on the next flight and, to my absolute amazement, he was right. I never told anyone this secret until now!

But even this trick won't help when the guns jam. Of course, jammed guns are never an excuse. In fact, in aerial gunnery, there are no excuses. That's why it is such a pure exercise in aeronautical skills. There are no excuses. Just as in aerial combat . . . there is no such thing as second place. You either win or you die. Life should be that simple!

But on this particular flight, I was robbed of my advantage when both of my two loaded guns jammed after just a few rounds. I panned after the first run and there were no holes. Damn it, I thought, I'm going to get stuck with a virgin banner. I turned off my master armament switch and continued to make "dry" runs, instructing my students on their techniques. But, the fun was gone out of the flight since I knew I would buy the beers and catch the heat as well. Each time I passed the banner, I panned and could see that my students were really hitting it.

As the flight approached the end of the range, a brilliant idea came to me. The tow pilot called the range "cold" as he left the Chocolate Mountain gunnery range and turned toward El Centro. It was then that I did an unprecedented thing. I passed the flight lead to the number-two man, and elected to escort the tow back to El Centro myself. Instructors never did that. It was boring and onerous . . . cruising along beside a banner just to keep other airplanes from inadvertently flying through the tow cable . . . or the banner, for that matter.

My flight returned to the field and I loafed along at 180 knots keeping station with the banner just 100 feet away. I could see plenty of holes. It was easily going to be a "century" banner (one hundred plus holes). Since I was the one shooting unpainted bullets (one of the colors was, of course, "plain"), I believed I could pull off the caper I had in mind to avoid bringing home a "virgin" banner.

I moved my airplane in so close that my wingtip momentarily touched the banner itself. I checked to be sure there were no farmhouses below us. Then I opened the canopy and drew the .455 Webley revolver I always carried for survival purposes. The roar of 180 knots of wind in the cockpit was deafening. I had been careful not to leave anything loose in the cockpit, because 180 knots of wind blast would suck any loose items out and away in a microsecond. I was also careful about my movements in the cockpit. It would have been easy for an elbow or a hand to venture into the slipstream. The results were scary to contemplate. Carefully keeping the barrel of the revolver out of the slipstream, I aimed it with my right hand as I flew formation with my left hand on the control stick. I hadn't the remotest idea what a 180-knot crosswind would do to my bullet. At that point, I began to have second thoughts. Was this right? Wasn't I after all cheating on my students? Shouldn't I take my medicine like a man?

Another voice within me countered with, "Horseshit. You do whatever it takes!" After all, isn't this how it all got started in the first place? One airman in a World War I biplane, armed with a revolver, (probably a Webley as ancient as mine), shooting at his enemy airman as they passed one another? Was that not how aerial gunnery began? With that spurious rationalization buzzing around in my head I drew back the hammer with my thumb and squeezed the trigger.

What followed was an explosion so deafening that I thought it must have ruptured both my eardrums. The little bubble of relatively quiescent air inside the cockpit must have absorbed every bit of acoustic energy as it bounced off the solid wall represented by that 180-knot gale. My ears rang for a week, but were otherwise undamaged. I reseated the revolver in its holster and slid the canopy shut. There was no way, I reasoned, that my bullet could have missed the banner. I was certain that my tumbling .455 bullet would make a "plain" hole sufficiently large that I could claim it as my one hit. That was all I needed . . . one lousy hit.

After the banner was recovered, there was much hooting and jeering by my flight over the paucity of "plain" holes, and any number of pungent references to my aeronautical abilities. But I smiled as I looked for that Webley hit. It had to be there somewhere!

It was not! Where that random bullet went is something I will never know. But it never hit that banner. That afternoon, with my ears still ringing, I bought a round of beer for my Blue Flight amid great ribaldry. I paid more than they knew for my virgin banner!

# 10

## THE PHANTOM BOGEY

I remember the day as clearly as if it were yesterday. Walking out of the operations building at El Centro, where I had just filed a flight plan to Alameda, California, I saw the jet. As I stepped out of the air-conditioned building into the blast furnace of the Imperial Valley mid-August morning, I spotted the unfamiliar airplane. It was parked on the aircraft transient line and looked particularly forlorn and abandoned. There was a moment before I recognized it as a Lockheed F-80C, the first production jet airplane built by the United States. It first flew early in 1949 and, though superseded by several newer jets, was a mainstay during the Korean Conflict.

But there was something different about this airplane. Compared to our T-33s which had a large external fuel tank mounted permanently on each wing tip centerline, the F-80 Shooting Star had a much smaller, jettisonable tank suspended under the wing tip. The ability to jettison tanks was an important feature of the F-80's combat capability.

What was so different about this particular airplane was that there were no drop tanks at all. It made the F-80 look quite different. By current standards it was smaller and lighter, with less power than the Cougar I was flying at the time.

Without the tip tanks, however, I knew that the F-80 was quite a spritely performer. That was because there were only 420 gallons of internal fuel. It was an extremely short-legged airplane.

Since I had a few minutes to spare and being mildly curious, I stepped back into the darkness and pleasant coolness of the operations building and knocked on the door marked "Assistant Operations Officer." At the gruff, "Come in," I entered.

*Lockheed F-80 (Navy designation TO-1) in flight on October 11, 1949. (Official USMC photo)*

A dour-faced lieutenant commander looked up inquisitively, his impatience clearly showing on a deeply tanned and seamed face. I introduced myself, stuck out my hand and, as he took it, asked.

"Whose F-80 is that parked out there?"

"It belongs to the air station; arrived yesterday," he responded, not sounding particularly happy about the matter.

"Oh," I replied, genuinely astonished. I knew that most, if not all, of the half dozen station pilots were propeller pilots, not qualified for jets. "Who is flying it?" I pressed.

"Nobody," was his answer. "Those God-damned idiots at AIRPAC just assigned it to us for proficiency flying. They didn't even bother to see if any of us is checked out in jets." The sheer delight I felt must have begun to show on my face because he appeared to become even more annoyed . . . especially at my next question.

"Do you mind if I fly it, just a little, until you guys get checked out?" I asked. I knew full well that they would never get checked out. There was neither travel money nor inclination on the part of the station's commanding officer to send any of his scarce pilots all the way to the jet transition training unit (JTTU) in Kansas. Even had he wanted to do so, the scarce training slots available would never be allotted to such a low priority as NAAS El Centro station pilots.

Noticing the FAGU patch sewn on my flight suit, he asked with a whine in his voice: "Don't you hot shots at FAGU get enough flight time without having to steal it from us poor station bastards?"

"I just want to keep it in flying shape until you guys get checked out," I lied.

"Okay, okay, God-damn it," he growled, handing a flight handbook to me. "But you are going to earn your flight time. Read the handbook, check yourself out, and then write me a set of open- and closed-book examination questions to start out my jet introduction training program."

"You got it," I responded, grabbing the handbook as if it were the Kohinoor diamond. I was out the door before he could change his mind. This, I decided, was going to be my secret airplane. No one across the field at FAGU was going to know about my good deal . . . nobody. It would be my own personal toy.

The flight to Alameda and back was uneventful . . . but I spent a good part of the long flight back planning how I would play with my new toy.

That weekend I burned the midnight oil and completed reading the handbook and memorizing the flight and emergency procedures section. I even made up a set of kneeboard cards to jog my memory on starting procedures, flight restrictions, recommended speeds and altitudes for the various flight profiles as well as emergency procedures. I knew the Assistant Operations Officer at the station would want such a set of cards anyway to include in his training syllabus . . . never mind that he would never get to put it into practice. That was his problem, not mine. The open- and closed-book exams were a piece of cake. It only took me two hours to work them up. Now, I was getting eager to strap this little machine on for size!

Tuesday afternoon, 14 August 1956. I sneaked out of the FAGU ready room after my second gunnery flight, informing the duty officer that I had some errands to run over on the station for a couple of hours.

I hid my flight helmet in the trunk of my car and headed for destiny, tingling with excitement. There is something thrilling about checking out in a different airplane, no matter what it is.

The F-80 was hardly a new airplane, battered and dirty-looking and still with its Air Force markings. The newly assigned plane captain walked around the airplane with me. We were both new at the servicing and maintenance procedures for the strange airplane, and I wanted to take extra precautions to be sure everything was right. We checked the fluid levels in the fuel tanks, the engine oil, hydraulic and pneumatic reservoirs. We checked tire pressures, shock struts and a raft of other details, any of which can get you in trouble if they are not maintained correctly. I even noted that the eager plane captain had filled up the water injection tank. However, since the handbook stated that this system was only to be used for operational necessity, I decided not to use it on my first flight.

In theory, water injection was a clever way to increase the thrust of a jet engine for a short time. By injecting a fine spray of water into the combustion chamber, the mass rate of flow of the engine was increased, thereby increasing its thrust. I decided to try it on a subsequent flight. The decision not to try it on my first flight proved to be a very prudent one.

When I finally climbed into the cockpit, I learned a few startling things, one of which was that it was too small. The other problem was the lack of a vertical adjustment for the ejection seat. If I sat fully erect, I couldn't close the canopy. As it slid forward, the canopy bow struck me in the back of my helmet a full two inches below its top. By scrunching down and leaning forward, I could close the canopy. After it was fully closed and I straightened back up, I found that my torso was still half an inch too long. Only by loosening my harness and sliding my hips forward and slouching could I raise my head fully. I made a mental note to manually jettison the canopy if I ever had to use the ejection seat.

My first flight in the F-80 was a pleasant surprise. It was astonishingly agile without the drop tanks and flew like a dream. It had a good roll rate and climbed like a homesick angel. The downside, of course, was that after 40 minutes the airplane was out of fuel.

The high point of the flight came when I "bounced" the four-plane flight of checker-tailed FJ-3 Furies as they came back from the gunnery range. All FAGU flights were considered "fair game" during the en route portions of their flights to and from the targets and ranges. They were expected to fly those portions in a combat spread formation and maintain a good lookout doctrine. If attacked, they were expected to engage the attacker, target times and fuel permitting. I circled the southwestern corridor of the gunnery range and listened on the radio frequencies which they used on the range. I over heard Lieutenant Colonel Bill Higgins, our Marine Corps instructor, call "off the range." I knew they were low on fuel and would be rendezvousing for their return flight to El Centro. They were dead meat as I rolled in from a position "up sun" and 5,000 feet above. Their number-three man was the first to spot me, but it was too late.

"Red One, this is Three, bogey, four o'clock high. Break right, hard, now," came the excited voice of the man in the second section. He was far enough abeam to represent no immediate threat to me so I pressed the attack on Bill Higgins' wingman. Even with a 50-knot speed advantage, I had no difficulty turning inside the defensive turn of the Furies. I closed to about 300 feet simulated a gun tracking attack, then executed a low "yo yo" maneuver which placed me at Bill Higginns' six o'clock for a gun kill on him, too. The number three man called out at that point.

"Red One, this is Three. I'm at your three o'clock and rolling in on the bogey." I looked over, saw him start in on me and broke hard into him, meeting him head on. As we passed, canopy to canopy, we broke into each other. He was still too slow and I knew, with my better turning performance, I could have him in two turns. Bill Higgins saved his bacon.

"Red Flight. This is Red One. Knock it off," came Bill's stern directive. They were low on fuel and had to retire . . . but, so also was I. I avoided making any radio transmissions because I knew they would recognize my voice. My secret would be out, and my good deal gone forever.

But there was suppressed rage in Bill's voice. He'd been "had" by an old F-80 and he was as mad as hell. The Furies joined up in echelon for their descent into the break at the field. Unable to resist the temptation to rub some salt into Bill's wounded ego, I made a high-speed pass close by his flight and pulled up into a victory roll before heading off toward the northeast as though returning to some other base. I was now really low on fuel, but I gave them a few minutes to dearm before I tried to sneak back into the field unnoticed. My engine was running on fumes, but my heart was filled with joy and satisfaction.

When I got back to the FAGU ready room it was buzzing with discussions about the phantom bogey over Chocolate Mountain. I feigned mild interest over noisy expostulations from Higgins' number-three man, who observed that the "wild-assed son of a bitch" had Bill Higgins and his wingman in less than 30 seconds, and was about to hand him "his ass on a platter" when Higgins called a "knock it off." "And then," he announced in a loud voice, (much to his leader's discomfort), "that arrogant bastard had to stick it up our ass even further with a frigging victory roll." It was all I could do to keep my composure.

I waited until the following week to fly my toy again. This time when I opened up the panel on the water injection tank filler point, I saw that it was full. The eager plane captain explained to me that he had just "topped it off." Curious as to where he was getting the water and how he was filling the tank, I stuck my finger below the surface and tasted the liquid. What I tasted made the hair stand up on the back of my neck. It was jet fuel!

The plane captain had filled the water injection reservoir with raw jet fuel! I shudder to this day at what would have happened if I had actuated the water injection toggle switch. Certainly there would have been a roaring engine fire and a melted turbine followed by an ejection in rapid sequence. For all I know, the engine might have exploded.

There was an immediate "come-to-Jesus meeting" with the plane captain. We drained the tank and purged it out with plenty of fresh water. Then I had the toggle switch in the cockpit safety-wired to the OFF position with a hand lettered placard nearby admonishing against its use under any circumstances.

My luck with the F-80 held out for several weeks before I was finally caught. The other flights were all replications of the first one. Rumors were rampant as to who this phantom bogey was, and why he came down here to harass the Navy types at El Centro. It was after one of my attacks that the exasperated flight leader was debriefing the flight. He heard an airplane enter the break and casually looked out the window. What he saw was an F-80 with no drop tanks . . . a very distinctive configuration. A quick telephone call to the operations duty officer ended my good deal forever.

"What's the pilot's name in that F-80 that just landed," he asked. They were all waiting for me a few minutes later when I walked non-chalantly into the ready room. Oh well, good deals don't last forever!

# 11

## "SKEETER"

Skeeter Carson was a little guy. But he had more energy than any three people. And, he was a real live wire; gregarious, with a well-developed sense of humor and a lively imagination. The expression typically worn by Skeeter was almost cherubic; but, with the hint that there was something enormously funny about to happen, about which he would not confide.

He was a Naval Academy classmate of mine . . . class of '52. He was also an aviator whose main claim to fame, at the time of this incident, was that he had beaten the Skipper in a critical race. It was almost like the salesman who beat his customer in a golf game and lost an important account. In fact, Skeeter became famous for this feat.

Skeeter's squadron was one of the first in the Pacific Fleet to receive a brand new airplane, a supersonic, all-weather fighter called the F3H-2N Demon. At the time, the Demon was really hot stuff! That was because it had a pretty good air intercept radar, supposedly could shoot the radar-guided Sparrow missile had an after-burner and even could go supersonic in a slight dive.

Anyway, Carson's squadron had a Skipper who wanted to demonstrate the amazing capability of his new airplanes so he helped "gin up" the stunt in which Skeeter's performance brought him fame. The stunt was to launch from an aircraft carrier off the coast of California and fly all the way to St. Louis, Missouri, in a record-breaking time.

The details of this aeronautical achievement will not be recounted here because they were not all that earth-shaking, as it turned out. What started out as a simple four-plane cross-country flight ended up being a race. Of the four, only two arrived at St. Louis on schedule. The other two arrived later after encountering difficulties enroute. Of the two that completed the mission, Skeeter Carson was first and his Skipper second. We, his peers, never let him forget this gaffe. Any right-minded junior officer should have known enough to let his Skipper win the race!

This, then was the backdrop before which our paths crossed. I was a junior officer and an instructor at the aerial gunnery training unit at El Centro, and Skeeter was a junior officer and member of a new Demon squadron which was deploying to El Centro for weapons training.

Although I knew he was on the base, I hadn't yet made contact with him. It should be said that Skeeter and I were in different parts of the brigade of midshipmen at the Naval Academy and were not together in flight training either. We were acquaintances, not friends.

In the spring of 1957 I was leading a flight of four F9F-8 Cougars on the Chocolate Mountain gunnery range. We had just completed our pass down the western side of the range and were finishing up. I called off the range, announced it was "cold," safetied my armament switches, directed number four to escort the banner home and called for numbers two and three in the flight to rendezvous with me at 20,000 feet in a right-hand turn toward home base.

At precisely that moment, I heard Skeeter's unmistakable voice, several octaves higher than normal, announcing his problem.

"Mayday, mayday, mayday. This is (I forget the call sign). I am just southwest of El Centro with a flame-out and unable to get a relight. I am ejecting. Out."

That was it! It took me several seconds to react. Then I passed the lead to number two by hand signal (the radio had come alive with queries from every quarter), waved him goodbye while pointing to home base, rammed the throttle to the firewall and headed for the general area of Skeeter's call. It was Gillcrist to the rescue of his old classmate . . . vintage derring-do!

Gradually, my Cougar built up a pretty good head of steam as I accelerated and eased down to 10,000 feet. I recall deciding to look for him from 10,000 feet for starters because most standard procedures called for bailing out of a stricken jet fighter at 10,000 if possible. My route was more or less direct from the little town of Holtville, passing east of the airfield and pointed roughly toward Signal Mountain at the Mexican border southwest of the field.

About the time I passed south of the field, I spotted a tall column of black smoke which was the crash site of Skeeter's airplane. I was perhaps 10 miles short of the crash site, level at about 9,000 feet, doing 450 knots and searching below me for signs of Skeeter's parachute in the air when something in my center windscreen caught my attention.

I looked directly in front of me, through the center windscreen and what I saw almost froze my blood! It was Skeeter! There, directly in the path of the airplane, level with me was my classmate. He was waving his arms over his head wildly. In an instant I was past him and could see the expression of abject terror on his face. He could not have been more than a hundred feet to my left when I roared by. I bent the airplane around to the left and slowed to a gradual descending spiral, following Skeeter down and relaying the information to El Centro Tower on Guard channel that I had found him, where we were and that he appeared to be uninjured.

The amusing thing that occurred while I was setting up my orbit was that Skeeter had ignited a day rescue flare which was issuing forth a billowing cloud of orange smoke. Although this was not a recommended procedure (for fear of setting the nylon parachute on fire, I assume), Skeeter had apparently decided on it as an alternative to being run down by this idiot who was now circling him. With the orange smoke trail to guide the rescue helicopter which was now airborne (he had Skeeter in sight), it became obvious that my services were no longer needed. Being low on fuel, I went home.

It was later on the next day when Skeeter and I talked on the telephone. He was furious with me for scaring him nearly to death. The more he remonstrated, the harder I laughed. He had concluded from my reaction that I had deliberately "buzzed" him. Ever since then whenever we have met, he has brought the subject up and given me heat. No matter how hard I tried, I have never been able to convince him otherwise.

But I learned a few important lessons from the incident. One is that a person in a parachute is extremely difficult to see from another airplane. The other lesson, one which I later learned from personal experience, is that parachutes descend very slowly, and seem to take forever to get to the ground from any reasonable altitude.

I am not sure what lessons Skeeter drew from the experience, but one of them must have been that his classmate, Paul Gillcrist, is either a sadist or has extremely poor eyesight. The answer probably falls somewhere in between!

# 12

## "LUMP'S" LAST FLIGHT

The first of December was to begin my last month at the Fleet Air Gunnery Unit, Pacific. It would mark the end of one of the best tours of duty I ever had. I knew that. After all, where else could a Navy pilot count on at least two live-fire weapons flights per day, five days a week, year 'round? It had been a tremendous experience for me and I was sure that I would look back in regret at having to leave.

On December 4 of that year I led a bombing mission on the loft-bombing target just west of the air station and north of Plaster City. It was a good flight. The students had all done well and the youngest student, a lieutenant, achieved the highest average miss distance and won the honor of buying a round of beer at Friday's happy hour.

At about the time I was getting ready to start briefing the second flight, word was passed to me that the Commanding Officer wanted to see me and that it was important. Stepping into the ready room where several instructors and students were gathered shooting the breeze or playing acey-deucey, I spotted one of my Cougar instructors, Lieutenant (junior grade) Bob "Lump" Davis. Instructors flew in one type airplane. I was the senior Cougar instructor.

"'Lump,' can you take my flight?" I asked him. "The Skipper wants to see me, and I'm told it's urgent." The qualifying comment was added by way of apology. Instructors almost never gave up an opportunity to fly, especially on a weapons delivery flight.

"Sure," he answered cheerfully, walking over and taking the briefing card from my hand. He strode briskly into the briefing room where the three students were waiting. As he disappeared from view, I overheard him say in a loud voice.

"Okay, Blue Flight, we're going to do some serious bombing." It was the last time I ever saw "Lump" Davis.

He was my closest friend at FAGU. Bob arrived from one of the premier Pacific Fleet fighter squadrons with the reputation for being a "hot shot" pilot. The very best youngsters in the fleet squadrons were candidates for choice instructor duty at FAGU. Every year the Skipper polled the commanding officers of all the fleet squadrons for the best candidates and screened them personally, picking out the ones he thought would make the best instructors.

Bob Davis was one of the most pleasant and likeable young men I have ever known. He had a great sense of humor and was quick with the funny quip. He also had a line of jokes that wouldn't quit. He stood about five feet eleven inches tall, was well built, had a thick crop of midnight-black hair, dark eyes and a deeply tanned complexion. Bob was good-looking in a rugged sort of way.

He arrived after I had been at FAGU for about a year, and we took a liking to one another right away. Since the fleet squadron from which Bob reported was a Cougar squadron, he was assigned as a prospective Cougar, or Blue Flight, instructor under training. As such, he was required to go through the entire five-week ground school and flight syllabus just exactly as the regular students did. The "instructor-under-training" was expected to score high on every syllabus flight and in the ground school syllabus as well. If he didn't, he wouldn't be long in FAGU. Bob was a whiz in all phases of instructor training. He graduated with honors and was quickly absorbed into the Blue Flight as one of its best and most aggressive instructors.

As an example of his finely tuned sense of humor, he stopped by my house one morning to pick me up and take me to work. Since my house was right on Highway 80 and on his way to work, it was no imposition. Bob's vintage black Porsche sports convertible roared into my driveway right on time and threw a shower of gravel onto my lawn. That was Bob!

I jumped in and we pulled onto the highway, leaving another shower of gravel, a cloud of dust and the smell of hot rubber behind us. It was a six-mile run west on Highway 80 to the turn off access road to the base. We had covered half of that distance when we caught sight of the unmistakable black Cadillac limousine, a vintage classic automobile which belonged to our new Skipper, Commander John Butts. Neither Bob nor I had gotten to know him very well yet and, as a consequence, I was keeping a low profile. The old limo was cruising along at a stodgy 50 miles per hour and as we approached from behind on the two lane black top road Bob had to downshift and decelerate. Naturally, he did this with a great deal of revving and gunning of the engine so as to be sure that the driver knew we were there. We ended up riding his rear bumper as Bob periodically eased the car to the left to look for oncoming traffic.

It was obvious that we wanted to pass, but our low-slung sports car couldn't see around the limo very well. The two-lane road called for caution before attempting to pass. From the passenger seat, I couldn't see anything and was forced to rely on Bob's judgment as to when it was safe to pass. Relying on Bob's judgment was not one of those things which I did well!

But I had the uncomfortable feeling that he was about to do something zany, embarrassing, stupid, or perhaps all of the above. Bob hesitated a moment, craning his head to the left to see around the limo. I was becoming extremely uncomfortable because we were tailgating our commanding officer's car, and the distance between our bumpers was less that two feet. Bob seemed to be waiting.

Suddenly, he reached over and opened the glove compartment in front of me and pulled out an object. I couldn't tell what it was other than it was black and fuzzy. He pulled it over his head much to my surprise and lo, he was wearing a wig of long black scraggly hair. It looked absolutely awful . . . like some mangy hippie, or perhaps a Jesus freak or a derelict from down in the barrio.

He then gunned the little black sports car, swung it viciously across the highway centerline and began accelerating past the limo. To my horror, I saw an enormous 18-wheeler bearing down on us like a freight train. It was not more than 100 yards away and looked 40 feet high. I recall seeing the trucker's hand reach up for the lanyard on his horn as I involuntarily gasped and blurted out, "Jesus Christ, 'Lump.'" Then I looked to my right where Lump was looking and found myself in the cold, baleful, blue-eyed gaze of our new commanding officer. It occurred to me that Lump was in disguise, but I was not. I didn't have the presence of mind to cover my face with my hands. As Bob whipped the Porsche back into the right lane just inches ahead of the Skipper's front bumper, the bow wave of slipstream from the truck struck us full in the face, buffeting the little sports car. At the same time, the loud blare of the diesel horn went off and down-doppplered as it went roaring by.

Bob looked up at the Skipper with a broad grin and jauntily waved his hand. He was smart enough to accelerate to 80 miles per hour the rest of the way to the turn off road. All I wanted was to get out of sight as quickly as I could . . . and lie low. God knows what my new Skipper must be thinking of me! Bob laughed all the way to the base, the wind whipping through that awful mane of black, scraggly wig. As we approached the main gate, Bob whipped off the wig, deftly slid on his garrison cap and threw the sentry a snappy salute. The Skipper never mentioned the incident to either one of us . . . and I couldn't understand why!

After being at El Centro only six months, Bob Davis applied for a slot which was coming open in the Blue Angels. After a few weeks he received a letter from the commanding officer of the Navy's Flight Demonstration Team ordering him to report for an interview. It must have gone well, because about a month later Bob received notice that he had been accepted and would be ordered in December . . . during the winter training season.

The fourth of December was Bob's last flying day at FAGU before he went off to the Blue Angels and, in his view, glory and the girls. When I stepped into the ready room that day looking for somebody to take my flight, Bob had already closed out his aviator's flight log book with the organization. In fact, he was checking out with the squadron duty officer when he cheerfully agreed to take my flight. I never gave it another thought.

My meeting with the C.O. was, it turned out, not important enough to warrant taking me off the flight schedule. That was my viewpoint. Obviously, it was not the view held by the C.O. I was sitting in my ordnance office later on when the telephone rang.

"Paul, this is the squadron duty officer. 'Lump' just crashed. He didn't make it. You'd better get over here quickly." For several seconds I just sat there stunned, with a dead telephone receiver jammed against my ear! I couldn't believe it! That cheerful, grinning, happy-go-lucky friend who just an hour ago had agreed to take my flight no longer existed! My friend, "Lump" was gone . . . forever. He would never be a Blue Angel. He would never be anything!

The sun was just sinking behind the crest of the Laguna Mountains when I pulled my car up to a roped off area in a farmer's field at the edge of a dirt road about 10 miles southwest of the air station. There was a young sailor who had been assigned sentry duty to keep curiosity seekers away from the crash site. The accident investigation board had just finished their initial search of the area and had left for the night. The crash site was a hole in the ground about 50 feet across and 15 feet deep. There was nothing visible in the bottom of the hole but soft earth and water. Somewhere down there, very deep, were the remains of an airplane and my best friend!

It took a bulldozer, a backhoe and a crane all the next day to reach the debris of Lump's airplane. A Grumman engineering team calculated from the geometry of the hole and the depth of the wreckage that the airplane struck the ground in a near-vertical dive at a speed of about 450 knots. The engine was found at a depth of 40 feet. Some of the key parts of the flight control system were recovered and subjected to intense engineering analysis. But, efforts by the accident board to find the "Easter egg" were unsuccessful. That conclusive bit of evidence was never found to establish a cause of the accident. It was nearly a month later that the final accident report was submitted. The board listed the cause of the accident as "undetermined."

But that day as I stood at the edge of the hole, I experienced a terrible sense of loss and sorrow. What a great guy Bob Davis was! What a lousy way to go! He had everything going for him. He was riding the crest of a wave: a weapons delivery instructor at the elite Navy weapons school, a selectee for one of the two solo positions in the Blue Angels, a bright-eyed young man whose star was rising. What a tragic way for that budding young career to end!

Something caught my attention across the roped-off area of the crash site. The sailor on sentry duty was showing something in his hand to a young lady, a curiosity-seeker. I strolled quietly around the perimeter of the hole and came up behind the young man.

"What have you got, Sailor?" I asked. He jumped, startled at the unexpected sound of my voice. I had surprised him.

"This, Sir", he answered. "I just picked it up in the dirt a few minutes ago." I looked into the cup of his upturned palm. It held a severed finger; the fourth finger of someone's right hand. It had been severed as if by a scalpel at the top of the third joint, where it joins the palm. I was shocked and angry. Angry that someone would show such a thing to a young girl, obviously to impress her. I found myself wondering how long it had really been in the young man's possession.

"Give me that," I demanded brusquely. He put it in my hand. It felt cold but strangely human in the palm of my hand. I quickly wrapped it in my handkerchief, put the bundle in the pocket of my leather flight jacket and walked to my car. On the drive back to the field I had this funny feeling about the only thing left of my best friend wrapped in a handkerchief in my pocket.

I had just turned onto Highway 80 headed east, back toward the air station when suddenly on impulse I pulled the car off onto the shoulder of the road. With the engine idling, I fished the bundle out of my pocket and opened it up. For some reason it was not a grisly sight at all. It was all that was left of my best friend. I felt the warmth of tears on my cheeks as I stared down at the pathetic remnant of a human being. The sharp rap of wood on glass has startled me. Looking up. I found myself staring into the clear blue eyes of a California Highway Patrolman. I hadn't noticed his prowl car pull up behind me.

"Anything wrong, son?" the policeman asked. It was a kindly voice, but the eyes turned stony when they recognized what was in my hand.

"No, Officer," I answered. "This is all that's left of a good friend of mine . . . the pilot who crashed back there."

"What are you going to do with it?" he asked.

I told him I was taking it to the senior medical officer at the base for medical analysis. Maybe they could learn something about the pilot's physical condition by running tissue tests.

"Are you okay?" he pressed. I said I was okay, thanked him for his interest, put the car in gear and drove away. He followed me at a distance until I passed through the main gate of the base. Then I saw the patrol car do a U-turn and head back toward the main highway.

It took me years to get over a feeling of guilt. I wondered a thousand times whether it should have been my right little finger being shown as a souvenir to a stranger. If I had not asked Lump Davis to take my flight, what would have happened to the both of us? Would I have ended up at the bottom of that hole? I guess, in retrospect, it is best that we don't possess the prescience to answer such questions.

My last flight at FAGU was on the 25th of December. The entry is in my logbook, but I can't for the life of me remember why I was flying on Christmas Day. After the holiday I left Nancy and our two young babies with her mother in San Francisco and headed east to Patuxent River, Maryland, the home of the Navy's Air Test Center. It was going to be a new and exciting chapter in my life.

As soon as I found a place to live, I would send for my family. I felt good, fulfilled and expecting a challenge as I crossed this wonderful country one more time by car alone. The syllabus at the Naval test Pilot School would be tough and demanding. I knew that. But, I had wanted for several years to be a Navy test pilot. Now was my chance. Life was good to me!

# PART III

# TEST PILOT

*"Success comes not from doing what you want to do, or what you like to do.*
*Success comes from doing what you have to do."*

*– Frank Tyger*

*The Naval Air Test Center at Patuxent River, Maryland, is truly a beautiful location. "The land of pleasant living" is how some brochure once described it. I recall finishing up a tour of duty as C.O of a fleet fighter squadron. It was a wonderful job, but I had already been told my next tour of duty would be in the Pentagon. I dreaded the thought of spending two years in that dismal place. But, they said, a tour in "Fort Fumble" was a career necessity.*

*In desperation, I flew one of my squadron airplanes all the way from Miramar, near San Diego, to Patuxent River. I had asked for an audience with the two-star Admiral who ran the test center because I had heard that he was looking for the right person to fill a job there for which I thought I was well qualified as my second tour of duty at Patuxent River. Anything would be better, I thought, than a tour in the Pentagon. But a nice flying job at the test center . . . how delightful it sounded!*

*The Admiral turned out to be a pretty nice person . . . a naval officer who could tell the difference between what I thought I wanted to do and what he thought I ought to do.*

*Certainly he could get my orders changed to Patuxent River. All I had to do, he told me, was to tell my detailer that was what I wanted. Then he added cryptically. "This is the land of pleasant living!" I thought, at the time, that it was an odd comment.*

*The rest of the interview was spent exchanging pleasantries. As I took my leave, I thanked him for spending the time with me as well as for his counsel. He shook my hand and looked me in the eye, saying "Just remember, this is the land of pleasant living." This time, I knew exactly what he was telling me. He was advising me, in his own subtle way, to go to the*

Pentagon. He was right . . . and I never thanked him! As the Arabs are wont to say, "Man proposes. Allah disposes."

This part of the book contains stories which took place while I was stationed at the Naval Air Test Center.

*The author at the Carrier Branch Test Pilot/Flight Test Division, Naval Air Test Center, Patuxent River, Maryland in 1959. (Author's collection)*

# *13*

## "TINKERTOY"

All Navy carrier airplanes have official names to go along with their numerical designations. As soon as the fleet pilots get their hands on the new airplanes, they have usually settled on a nickname which stays with the airplane to the tomb. The nickname is usually affectionate, but always irreverent. Thus it was that the A-4 Skyhawk came to be called the "Tinker Toy" . . . a name that stuck.

The same genius who in 1945 produced the legendary Spad gave the Navy "Heinemann's Hotrod" in the early 1950s. Ed Heinemann had produced another in a long series of winners and the A-4 would probably be remembered best in the chronicles of naval aviation as the workhorse of the Vietnam War. The "Hotrod" for a short time was nicknamed the "Scooter," but that name was quickly replaced by "Tinker Toy" and for good reason.

Ed Heinemann built the A-4 around the Volkswagen principle: small, simple, cheap, as few moving parts as possible, well-engineered, easy to repair and inexpensive to operate.

The A-4D-1 arrived in the Pacific Fleet in 1956. Commander Zeke Cormier's squadron, Attack Squadron VA-113, was the first to receive this new airplane. It was the hottest attack airplane ever to hit the fleet. The empty weight of a clean A-4D-1 was less than 9,000 pounds. With the Wright J65-W-2 axial flow jet engine delivering 7,200 pounds of thrust, the airplane had a higher thrust-to-weight ratio than any other attack airplane in the U.S. inventory.

Heinemann set out to create another legend, and he did it! The A-4D-1 weighed about half the empty gross weight stipulated in the Navy specification and had a maximum speed 100 miles per hour faster than that same specification. The XA4D-

*VC-1 TA4B in 1979. (Official U.S. Navy photograph)*

1 flew its first flight on 22 June 1954. The benefits of Ed Heinemann's design concept compounded themselves. The very size of the airplane had a huge impact on its vulnerability. But the size helped it avoid other weight penalties. The 27 1/2-foot wingspan was so small that the heavy structural and weight penalty associated with folding the wing was avoided. The wings didn't have to be folded. Furthermore, since there wasn't a complicated wing-fold mechanism, the entire wing structure could be used for fuel. Thus, the fuel fraction of the A-4 could be dramatically higher than other comparable tactical airplanes.

The Tinker Toy was in production longer than any other tactical airplane in the navy's history. As might be expected, there were many modifications made to the basic A-4, which began to make a mark unequalled in naval aviation history.

The A-4D-1 (A-4A) was followed by the A-4D-2 (A-4B), which included provisions for the Bullpup missile, a navigation computer, an in-flight refueling probe and a dual flight control system.

There is a "rule" which I call the "Gillcrist Rule." The rule says the U.S. Navy carrier based airplanes grow in weight at the rate of one pound per day for the entire time they remain in operation. The A-4 followed this rule "to a tee."

The A-4D2N (A-4C) had the nose cone enlarged to accommodate new equipment to improve its all-weather capability. These equipments included a better automatic pilot, terrain clearance radar and a new low altitude bombing system. The U.S. Navy ordered 638 of these airplanes for both Navy and Marine Corps use. The next version of the "Tinker Toy" was the A-4E, and it was equipped with a more powerful engine rated at 8,500 pounds of thrust. The "Tinker Toy"'s empty weight had increased so much that it badly needed more thrust.

The new engine, the Pratt & Whitney J52-P-6A, was subsequently retrofitted into those A-4As, Bs and Cs still in operational service. The A-4E showed up on Yankee Station in Air Wing FIVE in 1965 on board USS TICONDEROGA. One squadron was equipped with the older A-4C and the other s2quadron had A-4Es. The substantial difference in performance was dramatically demonstrated in the strike on the Haiphong Highway bridge on 19 April 1966.

The next version of the "Tinker Toy" featured a dorsal hump on the fuselage just behind the cockpit which housed electronic countermeasures equipment. Designated the A-4F, this latest version of the "Tinker Toy" also featured another engine upgrade and was powered by the Pratt & Whitney J52-P-8A rated at 9,300 pounds of thrust. It was this production run which also produced a two-seat version called the TA-4F, which was put into the Naval Air Training Command as an advanced jet trainer. The TA-4J was a follow-on trainer version. In 1979 the last version of the Skyhawk to be put into the U.S. tactical inventory, the A-4M was bought for the U.S. Marine Corps. The A-4M featured, among other things, improvements in the attack weapons system and a drag parachute.

By far the most versatile strike airplane the Navy has ever produced, the venerable "Tinker Toy" was the main battery of the fleet up until 1968 when the more capable A-7 Corsair II showed up on Yankee Station. Because of its lighter weight the A-4 Skyhawk was retained on the smaller ESSEX-class carriers for the remainder of the Vietnam Conflict. The U.S. Navy could not have carried out the air war in Southeast Asia without this workhorse of the fleet. Two hundred and fifty-seven A-4s were lost in combat in Southeast Asia. It took a former fighter pilot, Lieutenant Commander T.R. Schwartz, to show the light attack community that the A-4 could be effective in aerial combat. "TR" shot down a North Vietnamese MiG-17 with air-to-ground rockets in his "Tinker Toy" during a raid on the airfield at Kep.

I was given the opportunity to check out in the Skyhawk in 1957 when it was still new in the fleet and the hottest Navy tactical airplane except for the afterburner equipped F11F-1 Tiger and the F4D Skyray. Zeke Cormier offered me a ride in one of his squadron's airplanes at Miramar.

Zeke, the former leader of the Blue Angels, had been a student of mine at FAGU before reporting to VA-113. He agreed to give me one ride in the new airplane. I drove over to San Diego one weekend. Zeke gave me a one-hour briefing on how to fly the new Skyhawk then assigned his operations officer to "chase" me in another A-4. On take-off the chase airplane developed mechanical problems and was forced to return to the field. The last thing he told me before returning was that "I was on my own." I spent the next hour in the aerial combat area over Borrego Springs doing aerobatics to my heart's content. The airplane flew like a dream. I fell in love with it!

I got to fly the Skyhawk again when I left El Centro a year later and went to the Navy's Air test center for a tour of duty as a test pilot. By this time, a newer model, the A4D-2, was being tested prior to fleet introduction. The project pilot for the stability and control testing was Commander Jake Ward. Jake offered me an opportunity to fly his airplane one day to get "a particular test point."

"It is a simple enough flight," Jake explained to me. "All I want is a time history of a high-speed pitch-up with the center of gravity as far aft as you can get it." He provided me with some kneeboard cards with the airplane's center of gravity, expressed in percentage of mean aerodynamic chord, plotted against aft fuselage fuel cell quantity. Jake's airplane had been modified to allow the pilot to control internal fuel quantity to achieve aft center of gravity conditions.

As instructed by Jake, I climbed the "Tinker Toy" to 48,000 feet and circled while I transferred internal fuel as required. That took about 20 minutes. The high-speed pitch-up was to be achieved by entering a dive from that altitude under full power. When the Mach meter needle stabilized at 0.98, and with the photopanel turned on I was to pull back sharply on the control stick and hold full aft stick for the duration of the pitch excursion. It sounded simple enough!

The test proceeded as planned and the maneuver (aft stick) was started descending through 40,000 feet. With the altimeter needle whirling like a top and the Skyhawk in a 45-degree dive, I hauled back sharply on the control stick and held it in my stomach.

The nose of the airplane started up in fairly heavy buffeting. Suddenly the airplane snapped to the left so violently that my helmet slammed against the side of the canopy. The airplane snapped inverted and entered a wildly gyrating, left-hand inverted spin. My head was jammed against the top of the canopy despite the fact that I had tightened down fairly hard on my seat belt straps. The g meter pegged at minus three and three quarters. The airplane completed one and a half turns while I held the control stick full aft wondering how long I really wanted to stand this level of physical abuse. Then it snapped violently into an upright spin to the right. I remember wondering why Jake never told me this would happen.

I held full aft stick for two more complete turns, then neutralized all flight controls. The airplane recovered after about five seconds. I eased back on the stick and achieved level flight at 24,000 feet. There was something strange about the way the airplane responded to flight control movement. I determined the reason for the weird gyrations after a brief visual inspection.

The right leading edge aerodynamic wing slat had extended during the pitch-up so violently that the slat track had broken, locking the device in the "out" position. The slats are supposed to float in and out depending on airplane angle of attack. The left hand slat had not been damaged and had returned to the "in" position as it was supposed to. This gave my airplane an asymmetric condition which was "squirrely," to say the least. But the airplane was controllable and I returned to land, having achieved, I thought, my test objective.

*VA-94 A4C Skyhawk landing aboard USS HANCOCK in the Tonkin Gulf, circa 1967. (Tailhook Photo Service)*

I was standing with the test engineers looking at the wiggly lines that the various needles made on graph paper recording the time history of the maneuver. The g meter reading went right off the top of the chart three seconds after the start of the maneuver. The engineer pointed to the spot where the asymmetric slat failure had occurred and showed from other readings that the resultant excursion was indeed a one-and-a-half-turn inverted spin. The g value where the trace went off the page was "eight!" The airplane pulled something greater than that value.

Unbeknownst to me, Jake had sidled up to the group and was standing behind me looking over my shoulder. The engineer remarked that the airplane was grounded because it had been overstressed and would require a structural inspection.

"That's okay," Jake said with an impish grin at me. "That was the last flight anyway. Why do you think I saved this flight for the last?"

# 14

## HELICOPTER ORIENTATION

He was a short, roly-poly looking guy and I liked him . . . even though he never knew it! He was a U.S. Army Warrant Officer and a helicopter pilot assigned to the Rotary Wing Branch of the Flight Test Division. I had a rather bad habit of referring to helicopter pilots as "rotorheads." I don't do that anymore! Simply because they don't zoom around at the speed of sound shouldn't denigrate either their airmanship abilities or the importance of their mission. Some 10 years after I left Patuxent River I would have reason to thank a helicopter crew for saving my life!

But as a brand new test pilot in 1960, I was young and cocky and not nearly as mellowed as I am now. My usual derogatory remarks did not go unnoticed in the Rotary Wing Branch.

On 13 December, a bitter cold snap hit the tidewater area, and the Patuxent River froze over. For those of us who lived in the Solomons Annex, and had to commute to work by boat, it was serious. The only other way to get to work and back was by road. The nearest bridge across the Patuxent River was Mechanicsville, 30 miles north. So, the total distance by road was 65 miles.

The LCU (large Landing Craft Utility) which normally only replaces the standard picket boats in bad weather, had made the early morning run, crunching through a thin layer of ice which was growing in the river edges. We all suspected, as the temperature continued to drop, that we were in for a real freeze before the day was out. By noon the ice in mid-river was three inches thick and growing rapidly. The word was put out that all further boat transportation to the Solomons Annex had been canceled. I was stuck!

The car which I kept on the base was good, as many others like it, only for getting me from the boat landing to work and back. It was an old 1936 Plymouth which would never make the road trip to the Solomons. The tires looked like racing slicks. I could never ask Nancy to make the round trip from the Solomons in our good car. She had her hands full with three infants to care for.

So I swallowed my pride, took a deep breath and called the roly-poly warrant officer in the Rotary Wing branch. "You wouldn't happen to have a spare helicopter and a pilot who could fly me to the Solomons, would you?" I asked, honey dripping from every word.

"Sure," he answered genially. "I'd be glad to do it!" I was astonished that it had been so easy. What I had expected was some lip and a hearty "Go to hell," or worse. "Meet me at the H-23C parked behind the hangar in 15 minutes." I was chortling to myself, imagining all of my Solomons neighbors making the long drive in vintage cars or even staying the night on "mainside" in the bachelor officer's quarters. Here I was, getting a VIP helicopter ride right to my doorstep!

Stuffing some unfinished paperwork into my briefcase and shrugging into my overcoat, I headed for the helicopter, wondering all the while how I had been so fortunate! Nancy would be surprised and pleased to see me arrive home early, and the kids would get a great kick out of seeing their father arrive via helicopter. I took the time to call home and alert them before heading for the helicopter at a dead run.

My rotund friend was waiting when I arrived at the aircraft, an enigmatic smile on his face. In my eagerness to get into the helicopter and out of the bitter cold, I ignored the subtle warning signs.

We finished a hurried pre-flight check of the machine and climbed in. My friend had reserved the left seat for me. It was a tiny, two-place machine . . . but big enough for my purposes. God, it was cold! The steam from my breath began to fog up the plexiglass canopy. I sat there waiting. Nothing happened. He just sat there. I waited a little longer. Still nothing happened. Then I looked at him, our eyes meeting, and asked, "What's wrong?"

"Nothing's wrong," he answered, an innocent expression on his face.

"Why aren't we doing anything?" I asked, a little bit of exasperation creeping into my question. It was like a dam bursting when he answered! All of the pent up outrage of accumulated snotty remarks from jet pilots over the years poured out of him like venom.

"I am not going to do anything but keep myself alive," he hissed. "You are so God-damned smart, you are going to fly yourself to the Solomons all by yourself. I'm not going to do anything but watch. The only time I will touch the controls will be to keep you from killing me. Now, smart ass, if you don't want to go home, I frankly don't give a damn!" I had no idea his skin was so thin! But I was determined to go home in this machine no matter what it took.

"Well, at least help me get it started," I pleaded.

"Like hell I will," came the retort. "You can read, can't you?" He pointed to the engine starting instructions which were printed on a metal placard just below the throttle.

"Jesus Christ," I muttered half to myself as I started pushing in the plunger that primed the engine. The placard said, "Push four strokes for standard day; six strokes for a cold day." I pushed the plunger seven times, one for good measure and for the fact that I was freezing my ass off . . . this was an extra-cold day. My silent mentor never said a thing.

Opening the throttle about one inch, I hit the starter switch and held it while the starter whined laboriously and began turning the main rotor. I counted the requisite four blades (stipulated on the placard) and threw the magneto switch to the "BOTH" position (also stipulated on the placard). The engine coughed once or twice and started. Step one, I said smugly to myself. The placard suggested letting the engine warm up for a few minutes. I did so, trying frantically to remember any details from the single lecture we had received at the Test Pilot School on "Principles of Rotary Wing Aerodynamics." I remembered only dimly the mention of strange things like collectives, cyclics and blade tracking. It took me several minutes before I figured how to get the cabin heater going. My fingers by now were numb from the bitter cold. My mentor, I noticed, was heavily bundled up in a "Little America" – vintage sheepskin flying jacket and heavy mittens. Finally the cockpit began to warm up a little. For that small blessing I was grateful.

Now I was genuinely concerned over how in the world I was going to get this machine off the ground, much less fly it to the Solomons. My round friend gratuitously volunteered to make all the radio transmissions if I wanted. I nodded my head eagerly. Apparently I had surprised him by merely getting the engine started; concluding, I hope that I wasn't a total cretin after all!

Finally, since I couldn't think of any other pre-take-off things to do, I gave the plane captain the signal to pull the wheel chocks. When the sailor had done so and walked clear, I opened the throttle to takeoff power (also printed on the placard) and eased up slowly on the collective, hoping against hope that I wouldn't kill us both in the minutes that were ahead of us. His vow to keep me from killing us wasn't holding much water with me.

The helo's metal skids broke ground and we lifted off. I tried to stabilize the machine in a hover 10 feet off the ground as I had seen helos do in M.A.S.H. No way was that little machine going to allow me to do that!

What followed was something I hadn't anticipated. The little machine seemed to have a mind all its own. It started moving backward toward the hangar and descending toward the ground. My reaction was a near panic-stricken twisting of the throttle handle and a forward movement of the cyclic. The little marvel jumped upward and forward toward some high tension lines; I kicked rudder and moved the cyclic to miss them. The machine spun around like a dervish and we ended up heading in a direction I hadn't planned on at all. A loud noise broke through my panic. I looked over and saw my roly-poly friend doubled over in paroxysms of laughter. He was literally holding his sides while tears ran down his jowls. I didn't think it was funny at all! I thought we were in imminent danger of being killed at any moment! Finally, in desperation, I booted the rudder until the machine was headed for the Solomons and, cranking on full power, hauled up on the collective and "went for the moon."

The little helicopter wove drunkenly across the field, climbing roughly to the northwest, while my friend hurriedly made the necessary radio calls to accommodate for what I was apparently attempting to do in terms of altitude and airspeed.

The flight to the Solomons took only a few minutes once I had managed to level off at about 500 feet and stabilize on the right course and speed. Now I began to worry about landing the machine. I instinctively knew that it would be measurably harder than the takeoff. Managing to locate the large snow-covered field next to my house I set myself up for what I thought made a sensible approach. I even tried to figure out which way the wind was blowing.

Although the field was a good 100 yards on a side, it began to shrink in size as I approached it and tried to stabilize in a vertical descent. I was right. Landings in a helicopter are harder than takeoffs! I made it roughly through the transition to a hover about 50 feet off the ground, then a lurching, drunken, waltzing, out-of-control motion began which I simply couldn't dampen. My friend was again laughing himself sick. I got the machine finally under some semblance of control and we started down. A small cloud of snow rose to occlude all reference to the ground, so I did the only thing I could think of. I lowered the collective, hoping we wouldn't both die in the next few seconds.

Ground impact was pretty hard, but my friend didn't apparently think it was so bad. He let out a war whoop of satisfaction (and probably relief). I dumped the collective and throttled back to idle as my roly-poly mentor, for the first time took the controls. He was smiling . . . and so also was I!

I unstrapped, climbed out of the machine and turned toward the warrant officer. Before I could say anything he grinned and shouted, "Not too bad. I guess that makes you a rotor-head too, smart ass!"

I reached out and shook his extended hand.

"I'll never use that expression again," I shouted back at him! And I never did! He waved and took off for Patuxent River.

The kids, watching from the bedroom window, thought it was all very exciting! They hadn't the foggiest idea just how exciting it had really been!

# 15

## SUPER PERFORMANCE FIGHTER

My first project as a new test pilot was one called the Super Performance Fighter . . . and it was an exciting project. It involved the performance testing of a rocket engine intended for the North American A-5 Vigilante. The airplane was already in production and was exceeding all performance expectations as the Navy's new Mach 2 heavy attack airplane. Follow-on plans for the Vigilante included configuring it for a high altitude dash capability for its nuclear attack mission.

To achieve the high altitude Mach 3 dash capability, the A-5 would have a rocket engine mounted in the weapons tunnel between the engines. Just forward of the rocket engine (in the tunnel) would be a fuel cell carrying hydrogen peroxide fuel for the rocket engine. Hydrogen peroxide would be mixed with jet fuel from one of the airplane's internal fuel tanks to provide the proper fuel mixture for the rocket. Just forward of the fuel cell would be carried the nuclear weapon. Delivery of the weapon would include extrusion (ejection from the tunnel) of first the rocket engine, then the empty fuel cell and, finally the weapon. The whole concept was great, but the program never came to fruition.

Two North American FJ-4 Furies were modified as test beds to evaluate the performance of the rocket engine. They were designated XFJ-4Fs. This evaluation began while I was still a student at the test Pilot School. The initial ground tests were quite noticeable. Regardless of where I was on the base, I could always tell because the rocket engine made such an odd and deafening noise. Watching it was spectacular. A long sheet of flame would streak out behind the test bed airplane for a good 20 feet. Every few feet in that streak could be seen the diamond pattern of the standing supersonic shock waves. Whenever I watched or heard that engine, I

*FJ4-B (USMC El Toro Museum photo)*

found myself wishing that I could, one day, sit in the cockpit of an airplane with that monster strapped on behind me.

One of the test bed airplanes was assigned to Flight Test and the other to Service Test Division. The tests were almost completed when I graduated from Test Pilot school. But, shortly after I graduated, the test bed program was expanded to include a modified rocket engine that was throttleable. That was the project I inherited as my first project when I arrived at Flight Test.

The test bed airplanes had been extensively modified for the rocket tests. The rocket engine was mounted in a housing directly above the airplane's tailpipe in the base of the vertical stabilizer. The after fuselage fuel cell was re-plumbed and became the hydrogen peroxide fuel cell. When the rocket engine was in operation, hydrogen peroxide was pumped under high pressure through a silver catalytic screen which caused chemical ignition. The resulting combustion gasses were vented into a combustion chamber where they were mixed with jet fuel and then through a convergent/divergent supersonic nozzle.

Because of the high Mach numbers anticipated, the tailpipe of the Fury was modified to increase its cross-section. As a consequence, during normal, basic engine operation (without the rocket ignited), a device was employed to reduce the tailpipe cross-section to get adequate thrust to operate. This device was about as Rube Goldberg a creation as I have ever seen. It was a rough piece of three quarter-inch boiler plate attached to the bottom of the tailpipe by a hinge and normally lying flat against the bottom horizontally. Whenever the pilot needed to decrease the tailpipe cross-sectional area (for non-rocket operations), he simply pulled on a lanyard which pulled a cable which in turn caused the boiler plate to rotate 90-degrees on its hinge and extend into the tailpipe. It was crude, but it worked!

*This page and opposite: FJ4-F.*

Another modification to the FJ-4F was to the flight control system. The basic airplane had no stabilization augmentation systems except a fairly crude yaw stabilization system. A pitch stabilization system was added to give the airplane more controllability during its frequent and rapid excursions into and out of supersonic flight. That also worked. The other modification to the basic FJ-4 was to its environmental control system. Features were added to the system to allow the pilot to use what was the greatest abortion ever developed to protect pilots from the effects of super high-altitude flight . . . the partial pressure suit. It was absolutely horrible, extremely uncomfortable and very limiting in terms of the pilot's mobility in the cockpit.

The performance of the test beds with the rocket ignited was absolutely spectacular. It took a high subsonic airplane which could only go slightly supersonic in a dive, and turned it into "the king of the hill!" The down-side was that although the airplane could now out accelerate anything flying at the time, it could only do so for two minutes. But during that two minutes the FJ-4F was an airplane to behold. To be sure, the rocket installation in the Vigilante would not have such limitations. The two minute constraint was a limitation imposed by the size of the hydrogen peroxide cell in the FJ-4F. It would be substantially larger in the production Vigilante. Nevertheless, the throttling of the rocket engine became an important part of the development program because there was considerable disagreement among rocket scientists as to whether there would be greater operational utility in having an engine that could be throttled. Their estimates that full throttling of the engine would increase the burn time by a factor of three and extend the FJ-4Fs test time to six minutes. It was estimated that the A-5, during its Mach 3 dash at 70,000 feet, would be more efficient and flexible if it had an engine with a throttle.

The two-minute fighter became a standing joke around the bar at Patuxent River. For two minutes, the airplane could out-climb, out-turn and out-accelerate anything in the skies over Patuxent River. But at the end of two minutes, it turned

into a pumpkin. Woe to the FJ-4F pilot who put himself out on a limb in aerial combat when the transformation occurred. It could be downright embarrassing.

I was returning from a test flight one day in an F4D Skyray and spotted, directly over the field at 35,000 feet, an FJ-4 flying along fat dumb and happy. Clearly he was not looking out, I concluded, as I rolled into a simulated gunnery run on him. I tapped the Skyray's afterburner just a bit to get up a good head of steam and made my run from dead astern, pulling up directly (and closely) in front of him to give him a bit of jet wash to wake him up. Then I pulled up vertically and rolled inverted to look directly down on my victim in disdain. What I saw when I looked down was nothing but an enormous white smoke trail. I followed it with my eye and saw a tiny black speck in front of it. It was an FJ-4F (my victim) accelerating supersonic and a good 15,000 feet above me and opening like a freight train.

My "bounce" must have occurred just a fraction of a second before the rocket engine lit. After landing I called Service Test Division to see who it was that I had bounced and if I was in trouble for doing it. Wally Schirra came on the line and identified himself as the pilot. I asked him if he liked my bounce.

"What bounce?" he asked. "I never saw anything. I had my head inside the cockpit turning on the rocket engine. Let me tell you, when that thing lit off, that airplane shook like a son of a bitch before it took off!"

Whoever it was in the requirements business in the Pentagon who decided that the Vigilante's rocket engine had to be throttleable, ultimately killed the whole program. The test beds had been returned to Columbus, Ohio, for the modification work. When they were pronounced ready for pick, up I caught a flight to Columbus and showed up in my flight suit ready to deliver the first. As project officer, I had been invited to go to Columbus to observe the final test.

Excited about my first project, I arrived at the operations building in time to see the North American test pilot, Ed Gillespie, takeoff. Twenty minutes later I was standing in the flight operations office listening in on the flight test radio frequency. Ed's laconic voice announced that he was about to light off the rocket engine. Once ignition was complete and a good burn was established, he was going to throttle the engine back to about one half thrust. A few moments later he announced that he had a good light and a good burn, and that he was going to throttle back. There was a moment of silence then he shouted excitedly, "Jesus Christ, did you hear that down there? I think the whole ass end of the airplane blew up!" I've got control and the basic engine looks okay. But, I'm declaring an emergency and coming in for a precautionary landing."

Despite the loud blast which Ed heard, only minor damage had been done to the empennage of the airplane. But repair was necessary so I returned to Patuxent via airliner instead of my very own Superperformance Fighter. Needless to say, I was enormously disappointed. Two weeks later I returned to Columbus and there was a repeat of the same explosion. Shortly after the second incident, the program was canceled.

# *16*

## RIDE OF THE VALKYRIES

T he young sailor stood at my desk looking pale and distraught. He held in his hand a crumpled telegram from the Red Cross representative in his small hometown of western North Carolina. The telegram, which I had just returned to him, advised him that his father was terminally ill and was in the intensive care unit of the hospital near Seymour-Johnson Air Force Base. His father, the telegram further advised, was not expected to live for more than a few hours; and had specifically asked for his son's return.

Could I help? That was the simple request. I thought for a moment, considering the circumstances. It was the 4th of April, 1961. The afternoon was wearing on and it was raining torrentially. I knew that if I asked the boss, Captain Bob Elder, Director of Flight Test, he would approve. The problem was that it was Friday; and if I took the flight (who else would do it on such short notice?), I probably wouldn't return until early Saturday morning.

We lived across the river from the test center. The picket boats stopped running at 9 o'clock p.m. and didn't resume on Saturday morning until 10 o'clock. It would mean another one of those Friday afternoon telephone calls which always started out with, "Honey, I won't be home tonight . . ." I dreaded those calls and the ensuing string of questions, none of which I was very good at fielding. I also knew that I was going to take the young man standing before me to see his dying father. Picking up the telephone, I made the two calls.

Bob Elder approved the flight. My wife gracefully acknowledged my impending mission of mercy. She had been caring for three sick infants all day, and I knew the telephone call was a hard pill for her to swallow. I sent the sailor in search of a flight suit, helmet, gloves, boots and an oxygen mask. While he was doing that, I jumped in my car and drove to the operations building to get my weather briefing and file a flight plan.

The weather briefing was an unpleasant experience. The meteorologist consulted the bank of teletype machines behind him. They were busily clicking away, describing a picture of a line of thunderstorms 100 miles west of my destination and moving eastward at 20 knots . . . but accelerating.

"Sir," he announced forebodingly, "if this squall line stays at its present speed and direction, it will hit Seymour-Johnson about 40 minutes after your arrival. The terminal forecast, in that case, will be minimum visibility of one-half mile in heavy rain." I grunted disapproval, beginning to dislike the sound of the first shoe falling.

"However," he added, seeming to draw some pleasure from my pained expression. "If this weather continues to accelerate, the squall line is going to hit that field just about the time you arrive. And, Sir, it's going to be the damnedest frog-strangler you'll see in a long time. It will be a monster storm with gale force gusts, and they'll for sure shut down the field. You'd better be damned careful about the alternate base you pick . . . 'cause that's where you'll end up."

The T2V-1 Lockheed Shooting Star would have plenty of fuel when we got to Seymour-Johnson, so I picked my alternate field well out in front of the weather, Myrtle Beach Air Force Base, North Carolina, a good 120 miles east on the Atlantic coast. I was pretty sure that it was a safe bet. On an instrument flight plan, the Federal Aviation Agency rules require that a pilot include in his plan the route speed, time, altitude and fuel required to proceed to an alternate field in the event weather precludes his landing at the primary destination.

I did all these things and drove, now hurriedly, back to the Flight Test hangar to meet my passenger. I noticed that the rain had grown noticeably heavier. The windshield wipers weren't doing any good anymore and I had to slow down drastically, squinting warily through the streaked glass and fleeting fragments of clarity for the roadway.

The sailor was waiting nervously in the line shack when I sprinted across the parking ramp, getting totally soaked in the process. It was getting prematurely dark from the dense black cloudbank dumping on the tidelands like a giant waterfall. I took an extra 15 minutes showing the sailor how to hook his oxygen mask to his helmet and how the parachute and seat restraint buckles worked. I also gave him a few words of advice about the use of oxygen and the phenomena associated with high-altitude flight. Finally, I explained what he should do and what to expect in the event I should tell him to eject. It was a cursory bit of training, to be sure. However, I knew we couldn't do the usual routine of leaning over the edge of the cockpit explaining all of this in a downpour. It was absolutely essential that the canopy be open only long enough for us to step in and close the lid.

If very much rain were allowed to enter the cockpit, none of the electronic or electrical equipment would work. I would simply have to hope the young man strapped himself in and hooked up his oxygen and radio equipment the way I showed him.

Hurry as we might to get from the line shack to the airplane and then into the cockpit, we were soaked to the skin before the canopy finally spindled closed and the locks centered into place.

"How are you doing?" I asked my sodden passenger over the intercom.

"Okay, Sir," came the nervous response.

The engine start, taxi and clearance onto the duty runway were reasonably uneventful except for the fact that, without windshield wipers, the visibility through the plexiglass canopy was absolutely horrible. The ground controller had to give me a good deal of detailed guidance from his vantage point in the airfield control tower to get me to that point on the field.

The first 200 feet or so of takeoff roll were made on faith. All I could do was monitor my compass heading to keep from running off the runway. As soon as the airplane had 30 or 40 knots of ground speed, the airstream cleared the moisture from the windscreen and I could see where I was going. The tower personnel had even turned on the runway centerline lights early to assist me.

The flight to Seymour-Johnson was made at a cruising altitude of 35,000 feet. We were above the rainfall, but the turbulence in the dense black clouds was fairly heavy. It was pitch-black when we started our penetration to our destination. The turbulence and rain grew heavier as we descended through 10,000 feet. It was blacker than the inside of a closet.

Descending through 1,000 feet, we were informed by our radar final controller that the weather ceiling had lowered to 600 feet and one-half mile visibility in heavy precipitation with the conditions worsening by the minute. My thought at that point was that we had just slipped in, in the nick of time.

Descending through 400 feet on final approach the controller guided me skillfully to a point at which the stroboscopic approach centerline lights came into view directly in the center of the windscreen. On landing rollout a gust of wind swerved us violently and there was a sudden increase in the volume of pelting rain. For the next 15 minutes, the rain and winds were so violent that the ground controller told me to hold my position on the taxiway. The airfield was summarily closed as the squall line raced right across the field. Lightning flashes lit up a ferocious-looking black cloud rolling by.

Slowly the winds abated and the rain eased off. The surface winds shifted dramatically to signal the passage of the squall line and we made it to the transient parking area without further ado. After engine shutdown, we exited our airplane and ran to the operations building in light rain.

Because of the presence of lightning, I was told that there would be a delay in refueling my airplane. I overheard the sailor inquiring of the operations duty officer about bus schedules into town. At that time of night, I knew he'd have trouble getting to the hospital. There was a $20 bill in the small zipper pocket high on the left shoulder of my flight suit and it was soaking wet when I pulled it out. It was mad money . . . my emergency fund . . . all wadded up and damp with sweat and rain.

I walked over to him, thumped him on the back and pressed it into his hand. "Here, call a cab," I told him. "Time is too precious."

Tears came unabashedly to his eyes. I shook his hand and told him, "I'll say a prayer for your Dad. So long." I felt too embarrassed to hang around, so I turned on my heel and walked back to the line shack. I never saw him again.

Two hours later I taxied onto the duty runway for the return flight to Patuxent River, remembering the ominous warning of the Air Force meteorologist. He advised me that I had to fly across what was now a monstrous line of cumulo-nimbus clouds which ran roughly along my line of flight. He recommended that I try to cross them at a point 50 miles north of Seymour-Johnson where an airline pilot had just reported a relatively easy passage, with cloud tops at 40,000 feet. Since I had filed my flight plan at an altitude of 43,000 feet, I felt reasonably comfortable that all would be okay. But, he gave me a warning.

"Be careful," he had said. "These systems change by the minute. You may fly right into the big mother of the storm."

My plan was to cross the squall by flying over the top of it where our intrepid airline pilot had just crossed. Then I would parallel the system to the east all the way to my destination. It was going to be, in the vernacular of the trade, "a piece of cake!"

The windscreen of the T2V-1 began to clear itself of tiny raindrops as the airplane accelerated in its takeoff roll. The night was inky black. I focussed my attention on the dotted white-striped line painted on the centerline of the runway. I knew that if I let it get out of my view, either to the left or right, there would be no directional reference to steer by except my compass . . . and that was rough. A degree of error in the compass could easily run me off the side of the runway and into the proverbial ball of fire . . . an ignominious way to begin (and end) a flight.

Keeping the dotted line in sight became more difficult the faster the airplane went. Just the touch of the rudder would start the dotted line whizzing to one side or the other. Finally, to my relief, there was enough airspeed for me to rotate the nose and lift off. The main mounts skipped once or twice, with a little sideways lurch each time, and then I was airborne.

It was a relief to be airborne. Somehow I felt more in control of my destiny with my machine and me in the air, as dark, rainy and windy as it may be. The airplane's awkward transition to the "clean" configuration took eight or ten shuddering seconds while wheels and flaps laboriously tucked themselves into place and the airplane smoothly accelerated to climb speed. Finally the airplane, now in its proper milieu, began the long ascent into the gloom.

I knew, as we passed through 6,000 feet, that the bottom of the storm raging overhead, was perilously close. When I entered that maelstrom, I would knowingly be subjecting myself and my airplane to whatever abuse it was certain to heap upon us. All I could hope for was to get us safely through and on top of it as quickly as possible. Since my airplane could only claw its way up to about 46,000, feet I could only hope that the top of this monster storm did not extend above that level. If it did, I knew I was in for a ride!

The first clue that I was actually in the storm was the abrupt winking out of the few scattered lights on the ground below, the lights of a few farmhouses. Suddenly they were gone . . . snuffed out by the genie of the storm. Then came the soft drum of raindrops. Next, was the eerie blue halo of Saint Elmo's fire. The weird blue blanket of electric fire extended several inches from the skin of the airplane everywhere I looked . . . nose cone, wings and fuselage. Everywhere I could see my airplane there was this blue amoeba wrapped around me.

Finally, came the element that I dreaded the most . . . turbulence . . . it was inevitable! It began as a buzzing sensation, a light airframe vibration which grew steadily in intensity. Passing 25,000 feet, the first blow struck like a lightning bolt. Even though I knew it was coming I was nevertheless shocked by its violence.

One moment I was scanning the instrument panel, making small corrections with the control stick. The next moment I was flattened into my seat by a mighty blow equal to four or five time the force of gravity. The updraft into which my airplane and I had ventured hurled us skyward so fast that I had trouble reading the whirling altimeter needle. The next moment, the bottom seemed to drop out of the world. My helmet slammed up against the top of the canopy as I floated half an inch off my seat. The altimeter needle reversed its whirling direction and my frail airplane and I fell sickeningly into the black hole. For a full 5,000 feet we fell like a stone dropped down a well. Our crazy descent stopped abruptly with the seat being jammed up against my spine with tooth-jarring force. Again we began a dizzying ascent. I felt, quite simply, as though I no longer had control of my future.

The admonition of an old instrument flight instructor came to mind at this point.

"Someday," he had warned, "you're going to find yourself in the monster's mouth. When you do, remember that the only instrument of value will be your attitude gyro. Forget altitude and airspeed. Chasing them will be useless. Keep your wings level and your nose on the gyro horizon. Lower your seat. Turn up the cockpit lights as bright as you can get them. Ease a little throttle back from full military power . . . and ride the bastard out!"

I followed that pregnant advice. Easy as it may have sounded, it took all of my concentration. The wings rocked crazily. The nose pitched and dove as the airplane clawed its way through the horrendous pipe organ of up and down drafts. Each was a towering cylinder of fierce winds; either roaring dizzily skyward, or careening sickeningly earthward. The net effect of the two was a gradual ascent.

Passing wildly through 35,000 feet, with the rate of climb needle pegged at 6,000 feet per minute, I experienced the first lightning flash which totally destroyed what was left of my night vision. The wisdom of lowering my seat and turning up the lights became obvious. Never had I seen a flash of light so bright. I was temporarily blinded! Shooting pinwheels of fire obliterated the mute instrument panel. My mental clock ticked agonizingly and inexorably slow as I waited for my vision to clear. What would my attitude gyro tell me when it became once more visible? Would it show me upside down? Would I believe it? If so, what would I do?

Fortunately, when it once more became visible, the gyro showed the airplane in only a 30-degree bank to the left and the nose only 20 degrees below the horizon. No sooner had I leveled the wings and nose than the next lightning flash occurred. It was worse than the first! Again, I lost sight of the instrument panel. What would happen, I wondered, if one of these multi-million volt lightning bolts struck my airplane? I thought about the highly volatile fluid in the fuel cells and shuddered.

As the airplane laboriously ascended through 40,000, feet the violent thermals began slowly to abate. Somewhere in all of this activity the air route traffic controller queried me about turbulence. They were concerned, of course, about spilling martinis on airliners. Cocktails were the least of my worries, and I gave the controller short shrift.

I didn't want a sudden down draft to starve my gasping engine of air and snuff out the fire. An engine flame-out inside this monster could ruin my whole evening.

I was eternally grateful to note a momentary glimpse of a star-studded sky directly above me. Then, in an instant it was gone. My brief look had come from the bottom of a huge white bowl of towering clouds; then it was gone. Five minutes later my airplane and I emerged from the cloud tops into a breathtaking night sky. As far as I could see were the rounded tops of cumulo-nimbus clouds, each one with flickering genies of lightning captured inside.

The backdrop of jet-black sky and brilliant stars was magnificent! The flickering white domes resembled an array of giant Japanese lanterns at a galactic garden party. I leveled off tenderly at 46,000 feet and found that a line of cumulonimbus clouds stretched off to the northeast almost directly on a line with my route of flight to Patuxent River.

There was no moon, but the starlight was brilliant enough to read in the cockpit unassisted by internal lights. I toyed with the prudent tack of deviating either to the left or right of course and decided, to hell with that! I had clawed my way up through the monster . . . now I was going to ride her for all she was worth! Each time I nipped into the top of one of those flickering domes, the airplane would take a delirious dive into it, only to be spat out moments later from its side and into the clear. The exhilarating roller-coaster sequence repeated itself every few minutes for the rest of the trip to Patuxent River. It was more exciting than any ride I could have paid for at Coney Island! It was a storm I would never forget!

# 17

## TIGER

One of the most remarkable but least appreciated products of the Grumman Iron Works was the F11F Tiger. The airplane began in April 1953 as Model G-98, ordered by the Navy as a day, or clear air mass, fighter design contract. Clear air mass means being able to seek out an enemy airplane, day or night, but not in heavy weather. In other words, its weapons, the gun and the Sidewinder heat-seeking, air-to-air missile wouldn't work in the clouds.

Originally designated F9F-9, the airplane was a follow-on that would be the Grumman entry into the truly supersonic fighter community which began taking shape on the drawing boards of the aircraft industry designers in the early 1950s.

Production contracts were signed for the fighter and photo-reconnaissance models in 1954. The YF9F-9 prototype first flew on 30 July 1954, powered by a non-afterburning Wright J65-W-7 turbojet engine. In October of the same year, a second prototype fitted with an afterburner was rolled out. In January 1955, the second prototype began flying. The photo-reconnaissance production contract was canceled after the forty-second fighter model rolled off the production line. In April 1955, the Tiger's designation was changed F11F-1. A second production contract was let to produce 157 more of the fighter versions.

Initial deliveries of the F11F-1 were made in March 1957 to Attack Squadron ONE HUNDRED FIFTY-SIX (VA-156). A total of 12 Navy fighter squadrons, six in the Atlantic fleet and six in the Pacific Fleet, were equipped with the F11F-1. In April 1957, the Navy's Flight Demonstration Team turned in their F9F-8 Cougars for brand new F11F-1 Tigers.

On the aircraft carriers, a 12-plane F11F-1 squadron carried out a clear air mass role, while a 12-plane F3H-1 Demon squadron carried out the night/all-weather

*Grumman F11F-1 Tiger. (Grumman)*

fighter mission. During 1959, the F11F-1s began phasing out of the fleet and were sent to the training squadrons in the Advanced Training Command. The airplane was redesignated the F-11A in 1962.

I fell in love with the F11F-1 the first time I saw it perform at an air show at El Centro, California. After graduating from the Test Pilot School in the fall of 1958, one of my first projects was a performance and fuel consumption project on a Tiger configured with two external fuel tanks on the inboard wing stations. It was generally acknowledged in the fleet that the Tiger lacked sufficient internal fuel. To compensate for this deficiency, Grumman designers put internal fuel in every conceivable place in the airframe, even in the vertical tailfin.

The purpose of my project was to find out what effect the external fuel tanks had on aircraft flight performance . . . specifically, its range. I also used the F11F-1 to chase F8U-2 spin tests. Finally, I assisted in an F11F-1 spin project which had its interesting moments.

On one spin flight, for example, the entry conditions were intended to induce an inverted spin from a "hammerhead" stall. Entry was begun at an altitude of 30,000 feet and Mach 1.0 (the speed of sound), with a four g pull-up into a vertical climb. The throttle was brought to idle at about 40,000 feet and, decelerating through 80 knots, full forward stick and full right rudder were introduced simultaneously. The test plan called for holding these control inputs for three full rotations before neutralizing all flight controls to see if autorotation could be induced.

When I executed the maneuver, the airplane started into an inverted spin with a moderate yaw rate and a pitch excursion each 360 degrees of turn which became increasingly violent. Negative g went from minus 1 to minus 2 in the first turn. In the second turn the excursion was from minus 1 to minus 3.75, and on the third turn the g meter pegged at minus 4.

*Grumman F11F-1 Tiger. This aircraft was the author's project airplane, circa 1958. Note the experimental fuel tanks. (Official U.S. Navy photograph)*

Upon completion of the third turn, I neutralized controls and the airplane flipped into an upright spin, competed two more turns and recovered all by itself. During the playback of the cockpit data tapes, I heard myself saying something on the tapes which I do not recall having said. But when I said it, all of the engineers listening broke into gales of laughter. On the third turn during which the g meter pegged at four gs, we could all hear my labored breathing interrupted by a low moan and finally, "Jeeesusss Christ!"

I loved the airplane. It had a fine, natural feel to it and great control harmony. It was a dog in basic engine, so a good deal of afterburning was necessary in heavy aerial combat maneuvering against airplanes such as the Crusader and the Phantom II. In the landing configuration, it flew like a dream and turned out to be a fine carrier airplane except, of course for the fuel problem. The designers of the Tiger had produced a marvelously stable airframe with good positive, natural stability about the pitch and yaw axes. The only stabilization device in the flight control system was a simple yaw dampener, which hardly seemed necessary.

On 16 January 1957, Grumman rolled the first of two F11F-1Fs off the assembly line. They were powered with General Electric J79-GE-3A engines. The increased thrust gave the airplane a Mach 2.0 capability! The decision was made by the Navy to attempt to break the existing world's ballistic altitude record with this airplane.

On 16 April 1958 at Edwards Air Force, base the Super Tiger started its zoom climb at 40,000 feet with only 400 pounds of fuel remaining. The gutsy little fighter topped out at an altitude of 76,828 feet, breaking the existing record by over 6,000 feet. The previous record had been held by the British Canberra for over two years.

The F11F-1 was a small airplane, difficult to acquire visually in aerial combat maneuvering beyond a couple of miles. It had a clean, aerodynamic look and presented a very small profile head-on or tail-on. The airplane had a wing span of just under 32 feet. To improve the "deck spot", the outer three feet of each wing was hinged so that the wing tips could be manually folded straight down. It was a little under 42 feet long and weighed 13,428 pounds empty. The airplane held a little over 8,000 pounds of internal fuel which gave it a range of about 900 nautical miles. The fleet F11F-1 was powered by a Wright J65-W-18 turbojet engine equipped with an afterburner. The engine produced 7,400 pounds of thrust in basic engine and 11,000 pounds in afterburner. This engine gave the Tiger a maximum speed at sea level of 660 knots, Mach 1.4 at 38,000 feet and a service ceiling of 42,000 feet. The Tiger's armament included four fixed forward-firing 20mm cannons and it carried four Sidewinder missiles on four underwing stations.

It was a beautiful, simple, clean and honest airplane. But it was replaced by the bigger, faster and longer-range F8U Crusader.

# *18*

## A CLIMB TO SERVICE CEILING

It was 2 September 1960 when Gil Erb called and asked if I would like to go flying with him. Naturally, I said yes, without even asking particulars of the flight.

It was to be, I soon learned, a performance flight in a YAO-1 . . . a climb to determine the service ceiling of the airplane, just in case anybody wanted to know!

It was a weird looking airplane, to say the least . . . something like a bug in appearance. Also, it was an airplane under consideration for the U.S. Army as an artillery spotting airplane. My immediate thought was, who cares what the service ceiling is for an airplane designed to watch the impact of artillery rounds? Ceiling didn't seem like a relevant performance parameter for an airplane with that mission.

At the time, the U.S. Army didn't have a credible capability for the testing of fixed-wing airplanes. So they asked the U.S. Navy to do the testing; and the answer was yes. The test airplane was modified to carry a photo panel for the recording of flight data.

The YAO-1 Mohawk was a small, two crew, twin-turboprop powered, twin-tailed, high-wing airplane, and it looked like a beetle.

Our mission, Gil explained, was simple. We would climb the airplane at full power, and scheduled airspeed until the rate of climb fell below 200 feet per minute. That was the definition of "service ceiling," the altitude above which the airplane could not sustain a rate of climb of 200 feet per minute. We would mark that altitude, and the number, reduced to standard day parameters, would represent the published service ceiling for the airplane.

*Grumman YAO-1 Mohawk. (Grumman)*

The first indicator that this would not be a normal, garden-variety flight occurred in the flight test maintenance control office. The preceding test flight yellow sheets all told about the faulty cockpit heating system. It was listed as an "up gripe" (a maintenance discrepancy which did not preclude using the airplane for flying). After all, the maintenance control chief asked, who needed any cockpit heat in an artillery spotter in September? So Gil and I signed off on the airplane and took it for a mission test flight.

During the pre-flight briefing, Gil indicated that we would swap flight duties during the test flight. He would fly the airplane from takeoff to an altitude of 5,000 feet, and I would act as co-pilot and also record back-up performance data on my kneepad in case the photo panel quit. The data card included boxes for recording altitude, airspeed, outside air temperature, engine rpm, turbine inlet temperature, torque, oil temperature and pressure for every 1,000 feet climbed. At 5,000 feet, Gil explained, He would relinquish pilot duties, turn them over to me, and he would record the back-up test data. At 10,000 feet we would switch back again and repeat the process until the airplane reached service ceiling. Since no one, not even the Grumman test pilots, had gotten the airplane very high, we didn't have a very good idea what the service ceiling really was.

I do recall that the inoperative cockpit heater didn't enter into the equation in our test planning, it being September and all.

The start-up and takeoff were normal, and at 5,000 feet I took over the pilot duties. It was not a very demanding flight. All the pilot had to do was hold 140 knots and keep out of trouble. To my surprise, the airplane was a sprightly performer and we went through 10,000 feet like it was nothing. As briefed, Gil took over. The only thing I noticed was that the cockpit temperature had dropped to 65 degrees fahrenheit. Not to worry, I thought.

Both Gil and I were surprised how easily the airplane whipped past 15,000 feet, and I took over first pilot duties once again. Cockpit temperature had dropped to 40 degrees, but we were relatively unconcerned about it. Passing 20,000 feet Gil took over again and I noted that the outside air temperature (equal to the cockpit temperature) had now become a comfort concern. It now read below freezing, 30 degrees.

Our breathing had, furthermore, caused a visibility problem because it formed a layer of frost on the inside of the forward windscreen. This was a minor, and temporary, problem we thought as we wiped the frost off with a few swipes of our gloves. I also remember thinking that it would have been nice if either of us had thought to bring along a flight jacket. But not to worry, we thought. It will all be over shortly, so we can descend to warmer temperatures. The rate of climb of the airplane had noticeably slowed.

Passing 30,000 feet I relinquished pilot duties gladly to Gil and noticed that I was devoting a major part of my time to scraping thick ice off the windscreen. My teeth were chattering, I was involuntarily shaking, and this flight had long since gone off the bottom of the "fun factor" scale!

The Mohawk clawed its way past 35,000 feet and I took over once again. My body was shaking so badly that I was no longer able to hold a steady 140-knot climb speed. After a few minutes of this, I turned to look at Gil, whose face was a pronounced blue around the oxygen mask, and asked the obvious question.

"What the hell are we doing this for? This is insanity!"

"You are absolutely right," he responded. "This a frigging artillery spotter . . . and we are on our way to the moon. We're knocking this silly climb off right now! Who gives a crap if it still climbs at this altitude?"

The descent was a heavenly return to bodily warmth, even though my core temperature wasn't back to normal for several hours after we landed.

Needless to say, the flight test engineers were irritated that we had elected to abort the test objective because of some silly concept of bodily discomfort. Who the hell did we think we were, abandoning the project like that? What ever happened to the legendary test pilots of old, when men were men of iron and flew fabric airplanes? Neither Gil nor I felt like dignifying such questions with an answer.

What to do? It had been the last test flight. The airplane was due to be returned the next day. What about the service ceiling? Gil's violent answer was "screw the service ceiling!"

The test center's final flight test report showed the service ceiling as displayed on an X-Y chart with a solid line extending to 36,500 feet. Beyond that, the line becomes dotted and levels off at an extrapolated 37,850 feet (corrected for test data). To my knowledge, no one has been foolish enough to check it out, not even with an operating cockpit heater! So much for iron pilots and fabric airplanes!

# PART IV

# THE SEAGOING BOOMERANGS

*"I wish to have no connection with any ship that does not sail fast,
for I intend to go in harm's way."*

*– John Paul Jones*

*We called ourselves the "World-Famous Seagoing Boomerangs" of Fighter Squadron SIXTY-TWO (VF-62). We were one of the two fighter squadrons in Air Wing TEN, embarked in USS SHANGRI LA (CVA-38). There was a wry irony associated with the epithet because, on SHANGRI LA's previous deployment to the Mediterranean Sea before I reported to the squadron, it had been off-loaded to Naval Station Rota, Spain, inside the Strait of Gibraltar for almost the entire tour.*

*Naturally, the remainder of the air wing, the ones who stayed on the ship, took umbrage with the "seagoing" part of our name. Of course, in grim retaliation, our squadron pilots "complained bitterly" over being left ashore with all that wine and all those pretty Spanish ladies.*

*If the truth were known, it was really a bitter pill for the squadron to swallow . . . to be left off the ship-air wing team. Fortunately, when I arrived, the tour ashore in Rota was history. Anyone who tried to rib me about that aspect of the squadron's history was told unceremoniously to "stuff it." We soon stopped hearing that kind of crap from the rest of the squadrons in the air wing. But, there was a lesson in all of that which I did not forget!*

*Ten years later, when as a wing commander I was approached with a similar suggestion for our deployment to the Mediterranean, the idea man who came up with it had his head nearly bitten off. His problem was that he never understood why . . . and never even tried to find out.*

*The lesson was that a good manager and leader should never tell a part of his organization that they are not really part of the team. At least, not if the leader ever expects to need them in times of emergency.*

*Every time my big-idea man would suggest off-loading a squadron, I would bite his head off. He had a perfectly flat learning curve!*

*I suppose, in retrospect, I should have taken the time to explain it to him. But, it was so much fun to see him continue to take that near-lethal dose of gas . . . time after time, that I allowed it to become one of my more innocent forms of entertainment on an otherwise frustrating Mediterranean deployment. Some naval authority once said:*

> *"When principle is involved,*
> *Be deaf to expediency."*

*So much for expediency!*

*The stories contained in this part of the book all occurred while I was a member of the "Seagoing Boomerangs." Also, during this period, VF-62 was temporarily shore-based at Naval Air Station Cecil Field, near Jacksonville, Florida. The Cuban Missile Crisis arose during this period resulting in the squadron being tasked to take part in simulated contingency operations training flights against Cuba.*

# 19

## TIMMUQUANA

Timmuquana is the name of the road! Somewhere west of town, the name changed to 103rd Street, and it continued, under that name, 12 long, lonely, two-lane, black-topped miles west from the city of Jacksonville to the main gate of one of the Navy's master jet bases, Cecil Field. In 1960, the two-lane road had no shoulders for a good bit of that 12 miles. Just beyond the edge of the black-top on each side was a shallow ditch.

It's only a guess, but I'd be willing to bet that more Navy pilots have died on that particular stretch of road than on any other in the world! Friday night, happy hour night at the Cecil Field Officer's Club, was the killer night. The combination of alcohol, a narrow road, darkness and too much speed was deadly.

It was not surprising then that two fighter pilots walked out of the club at about seven o'clock on a Friday night in October 1963 and headed for their cars. They were both in high spirits, laughing uproariously over a shared joke. The taller of the two walked with a pronounced limp in his left leg and had a cigar clamped in his mouth. The smaller one had reached the point in his evening at which almost everything seemed funny. His laughter was giddy and infectious.

From their general behavior, it was obvious that they were happy in their world of fast fighter planes, camaraderie, a good laugh, several beers and thoughts of a snug home and family just 12 miles ahead. They proceeded carefully, and in single file, at the posted 30 miles an hour on the long, straight road headed north from the club to the main gate. The smaller of the two, driving a sleek, black Ford sedan, was the formation leader. The one with the limp trailed in a fast, sleek-looking white Lincoln Capri.

*Flight of VF-62 F8B Crusaders of Air Group 10, USS SHANGRI-LA (CVA-38), May 1961. (Photo by S.E. Harrison)*

Their sedate pace ended as they filed out the gate, past the sentry and turned right onto 103rd Street headed east. There were no lights illuminating the road, and hardly any farm houses along the way to offer any illumination. An overcast sky added to the darkness. The lead car kicked up the speed to a respectable 75 miles per hour. The wingman hung right in there, even during the acceleration, keeping no more than 10 feet between autos.

About a mile down the mordant road, the rear car pulled left into the oncoming lane and crept up alongside the flight leader. The wingman pointed to the cigar in his mouth and made a lighting motion with his right hand. The formation leader glanced quickly to his left, saw the signal and nodded his understanding. He also noticed during the glance, that his wingman was intently watching his leader, and maintaining no more than six inches between cars.

It took total concentration, so much so that the wingman's entire attention was focused on the black sedan . . . and the inches separating them. The leader, a grin on his face, kept the Ford going down the road straight as an arrow, hugging the right edge of the blacktop road.

With his left hand he felt in the left breast pocket of his khaki uniform shirt, fumbling for a book of matches. Finding it, he blindly reached to his left, leaning as he did so, and extended the book of matches out his window and into the outstretched hand of his wingman. The leader never took his eyes off the road. The wingman never took his eyes off his leader. With the precious book of matches in his hand, the wingman leisurely eased off on the gas just enough to drop back in behind his leader, just as an 18-wheeler, seeming 40 feet high, went roaring by, diesel horn blaring, lights blinking high and low beam frantically. The horn dopplered by and the wingman occupied himself with lighting his cigar . . . taking care to maintain the requisite 10 feet (no more) between bumpers.

The grin on the leader's face froze as he realized that his wingman, having lit the cigar, was now moving back up to the abeam position, presumably to return the worthless matches by reversing the heart-rending sequence of a few moments earlier. There was another set of truck headlights bearing down on them. The lights were frantically flickering high beam to low and back again in warning. Catching the high beam action, the white Lincoln returned the signal. It took another agonizing, heart-stopping couple of groping seconds before the leader had his fingers wrapped around the fateful matches. It was obvious that the Lincoln was going to stay there until the transfer was complete. Again, just at the last second, the wingman dropped back and fell in behind. Again the klaxon horn and the down doppler and the blast of wind as the truck roared by.

By rights, the Timmuquana should have claimed another pair of lives that dark night, but failed to do so because of the consummate driving skill and sheer stupidity of the formation leader and his intrepid wingman!

# 20

## GAUNTLET

In recent years the word "gauntlet" has achieved some negative press from the Tailhook '91 fiasco.

However, in one of its principal definitions the word still stands for a glove such as that worn by a knight. In this context, "throwing down the gauntlet" still means a challenge to a duel, as in the day Commander Herk Camp threw down the gauntlet at Commander Billy Phillip's feet.

Lieutenant Commander Herk Camp was the executive officer of an F8 squadron at Cecil Field. He was a tall, trim, affable young man in his mid-30s, always with a smile on his face and a joke on his tongue. He enjoyed the reputation for being a very professional aviator . . . a no-nonsense kind of pilot who had done well in the flying business. He had little regard for the small group of elitist egomaniacs who, over the years, had managed to give the Crusader community a bad name.

So it came as no surprise that he ended up in a face-off with one of them at long last. It happened, as did so many other face-offs, at the Officers' Recreational Facility (ORF) on Friday afternoon at happy hour. Herk had heard enough of the typical macho "BS," and the several martinis he had consumed caused him to make an uncharitable remark in the general direction of a noisy group to whom Billy was holding forth with particular hyperbole. Billy took immediate umbrage, and the verbal preliminaries to fisticuffs began. Poor Billy, having consumed more alcohol than his opponent, was faring the worse for it in the exchange. Finally, Herk concluded with, "Billy, if you flew half as well as you talked, you'd be a pretty good pilot. The trouble is that you don't!"

This inflammatory remark to such an exalted member of the Crusader community brought a shocked hush to everyone within earshot.

"Maybe you think you're better, Herk," came Billy's thick retort. Billy continued now in a louder voice; loud enough for nearly everyone in the club to hear, "Tell you what, Herk, I'll meet you head-on at 20,000 feet 0800 Monday morning in the dogfight area and I'll have your ass in three turns. I'm going to put my month's pay down to show you where my mouth is!"

Without thought or hesitation, Herk roared back at him, "You're on!"

When word of the impending duel got out, all bets were on Billy. He was good; no question about that. He spent nearly all of his flying time teaching air combat maneuvering (ACM), as it came to be called, at the fleet replacement squadron, VF-174. It was expected to be no contest. The only cause for bets to be to be offered and accepted was the three-turn stipulation. That was considered a pretty decisive win between two experienced Crusader pilots. News of the impending shoot-out spread around the base like wildfire.

Monday morning came and the two Crusaders took-off as a section headed for the combat maneuvering area to the south and west of Cecil about 25 miles, with Herk Camp leading. Everyone on the base who was either not flying or getting ready to fly was clustered around their squadron base radios which were all tuned to the agreed-upon dogfight radio frequency channel.

The flight leveled off at 20,000 feet as specified in the bet as they entered the training area. Herk, who had put Billy on his left wing, was heard to order, "Splitting off." As briefed, Herk turned right and Billy turned left to establish an offset for the fight. The airplanes slowly separated until Herk decided they were far enough apart to meet head-on when they turned toward one another.

"Turning in," Herk called aloud as he turned hard left toward Billy, his throttle jammed to full afterburner. It was a classic beginning, with Billy turning right toward Herk, his throttle also in full afterburner. A hush fell over crowded ready rooms all over the base as listeners waited for the next call, when the two Crusaders called passing each other that the fight was on.

The next call was Billy's gravelly voice shouting, "Fight's on!" There was a pregnant silence of perhaps 10 or 12 seconds and then came the almost strangled, outraged voice of Billy over the radio. "You yellow-bellied, chicken-shit son of a bitch."

Herk's account, repeated at lightning speed around the base just 10 minutes later, is the only account ever recorded of the epic event. Herk had strategically positioned himself to the right and west of Billy for a good reason.

As the two airplanes turned toward one another, Herk was headed coincidentally, directly at Cecil Field while Billy was headed directly away from home base. The two Crusaders hurtled toward each other, each at nearly the speed of sound, and passed canopy to canopy in the classic beginning of a dogfight. Billy wrenched his Crusader into a shuddering eight g turn toward Herk, seeing his opponent disappear from sight down his left side and over his left shoulder. That part of the fight was standard.

The rest of the fight was definitely not standard, because Herk never turned! Instead, he "bunted" his airplane over into a negative g acceleration maneuver, went supersonic and pointed his Crusader directly at the break for the duty runway at Cecil Field. It took four or five seconds for Billy's airplane to turn enough to put Herk's airplane back into his field of view, but he was nowhere to be seen. After another four or five seconds of wrenching turn, Billy was able to pick up in the distance the tiny speck which represented Herk's airplane now clicking off 700 knots as it headed for Cecil Field. Herk was out of range with an opening velocity of over 300 knots. Within seconds, he would be touching down on the north-south runway. Billy had been "euchred" and he knew it. The sense of outrage in his radio transmission was justified. He knew that the price he would have to pay would make his lost month's wages pale to insignificance.

Billy never quite got over this experience.

# 21

## BRIDGECAP

A senior naval officer's bridgecap cost about $30.00 at the time of this event. This was several dollars more than the price of an almost identical cap worn by junior officers. The difference was the cost of the added "scrambled eggs" on the bill of the former's cap. But, there was at least one junior officer at NAS Cecil Field who thought $30.00 was a bargain.

Lieutenant Commander Joe Stanley was a maverick, a rebel and, most of all, an iconoclast. Joe was the operations officer (my counterpart) of one of the two light attack squadrons in Air Wing TEN. The two squadrons were, it seems, locked in a continual, fierce competition to prove which was best in almost any conceivable category, including high jinx.

Joe was clever and possessed a biting sense of humor. His practical jokes had long since become legendary in the fleet. I always pitied Joe's counterpart in the sister light attack squadron. The poor guy never had a chance.

The denouement of this competition occurred on a Friday afternoon at happy hour at Cecil Field's Officers' Recreational Facility (ORF). There were easily over 300 spirited naval aviators in the bar having a wonderful time when Joe's perceptive eye caught sight of the commanding officer of the sister squadron, Commander Stu Nelson, enter the bar with his bridgecap on!

Joe was quick to recognize the terrible gaffe which Stu had committed, and even quicker to capitalize on it. Pointing at Stu, Joe shouted in a loud voice, "Drinks on the house!" Unfortunately, the noise level was so high that no one heard him, and Stu, recognizing his error, beat a hasty retreat from the bar, disappearing from sight.

Since time immemorial it has been a rule that anyone who has the temerity to walk into a Navy officers' club bar with his hat on must buy drinks on the house. In fact, in every bar there is a ship's bell suspended from the ceiling which is supposed to be rung to announce such an event to all present. Stu knew the rules and must have decided it would cost him several hundred dollars to buy a round in that full house. Being a little penurious, Stu must have decided to get lost, hoping no one but Joe had witnessed his mistake. Stu had, once again, sadly underestimated Joe Stanley.

About 15 minutes went by before Stu reappeared quietly in a corner booth, joining a few friends for a drink and a game of liar's dice. Joe spotted him immediately and he called two of the squadron junior officers into a quieter corner of the bar for a strategy session. What, he posed, should be done about Stu Nelson's blatant disregard for such a hallowed navy tradition? The two henchmen, Vosseler and La Cagnina, as I recall, disappeared on a mission obviously conceived by Joe's fertile mind.

Now Joe Stanley had a particular abhorrence for pompous people, and never passed up an opportunity to embarrass them and let a little bit of air out of their sails. Here was his chance to embarrass the sister squadron (Joe referred to them as "brand X"). Stu Nelson's time had come and he didn't even know the clock was ticking.

"Ladies and gentlemen, may I have your attention, please!"

I recognized Joe's voice and looked up to see him standing on top of the bar with a microphone in his hand. Except for the few people nearby, Joe's odd behavior went largely unnoticed. However, after two or three more attempts people began to notice, the roar of the crowd began to abate, and Joe Stanley was once again orchestrating one of his gambits.

"Ladies and gentlemen," he repeated, "a few moments ago one of our more distinguished naval aviators walked into this room with his bridgecap on." Joe paused for the chorus of hoots and catcalls which this announcement evoked from the boisterous crowd. Joe continued, "The gentleman in question chose not to honor tradition and sneaked out, hoping he hadn't been spotted." Again, Joe paused for the crowd to voice the proper level of outrage, which it did. There were calls, among other things, to "lynch the bastard." Joe was obviously enjoying the moment to its fullest, and I could see poor Stu, now trapped in the corner, squirming uneasily. He had the look of a man ready to walk to the gas chamber.

"Now, I'm not going to mention the person's name other than to tell you that he is the commanding officer of brand X"! The hoots, catcalls and Bronx cheers were now deafening. "Ladies and gentlemen," Joe continued, "I now direct your attention to the top of the three-meter diving board in the pool."

Everyone in the room did a 180-degree turn and directed their gazes to the swimming pool clearly visible through the enormous window in the opposite wall of the room. The bar was on the second floor of the club and commanded a panoramic view of the pool.

There on the very end of the high diving board stood Joe's two henchmen, Vosseler and La Cagnina, holding down two very large, fully inflated meteorological weather balloons. Suspended between the balloons was an officer's bridgecap with the scrambled eggs clearly visible on its bill. Suddenly, everyone in the now-quiet room knew instinctively that the name inside the hat band of the cap was "Stu Nelson." When Joe sensed that the moment was right, he gave a papal flourish with his hand and the balloons were released. They ascended majestically carrying the cap aloft accompanied by a deafening ovation from the crowd.

Once the balloons cleared the treetop level, prevailing westerly winds carried the hapless cap eastward toward the Florida coast. Joe later contended that an airline pilot reported two weather balloons, carrying an unidentified object, 50 miles east of Jacksonville at an altitude of 37,000 feet headed for Bermuda. Stu never forgave Joe for the loss of his cap and the embarrassment associated with the incident. But in retrospect, it was a great deal cheaper to replace a $30.00 bridge cap than it would have been to pay that $300.00 bar tab!

# 22

## DAWN PATROL

During the Cuban missile crisis in the fall of 1962, extensive preparations were made to execute an elaborate plan whereby U.S. tactical aircraft stationed in continental U.S. (CONUS) struck at pre-selected targets in Cuba.

Whenever an East Coast Navy carrier air wing returned from a deployment to the Mediterranean, it assumed some responsibility for a part of this elaborate master plan. This, of course, meant an enormous amount of strike planning had to be done by those individual pilots and strike groups to ensure they could make it to their designated targets, hit them and make it home safely. The enemy territory had to be learned. Their air, electronic and air defense orders of battle had to be memorized. Routes to and from targets had to be carefully laid out to ensure they skirted the many heavy concentrations of anti-aircraft artillery, surface-to-air missiles and fighter airfields as well.

Finally, the question of blue-on-blue (friendly-versus-friendly) attrition became a serious concern. There were so many U.S. strike airplanes transiting to and from targets in Cuba that conflict among them became almost as serious a threat as that posed by Cuban fighters and surface-to-air weaponry. This whole drill was strangely prophetic of the problems which were experienced in the Persian Gulf War almost thirty years later.

As a member of Fighter Squadron SIXTY-TWO, (VF-62) an element of Air Wing TEN recently returned from a Sixth Fleet deployment, we were deeply involved in this strike planning effort. A series of air wing strike cells were formed and targets were assigned to individual cells. Our cell was six airplanes, four A-4C Skyhawks and two F-8B Crusaders. The A-4s were loaded with external fuel tanks and Mark 83 one-thousand pound bombs. The F-8s were loaded for a fighter escort mission with two Sidewinders and full 20mm ammunition.

After we had completed our planning, we concluded that only if everything went exactly as planned and there were no "glitches" would we ever make it back to Cecil. I personally envisioned our making it back only as far as maybe Homestead Air Force Base, or worse yet, NAS Key West; that is if the flak, SAMs and MiGs didn't get us over Cuba. The risk element associated with taking out our assigned surface-to-air missile site in eastern Cuba was not inconsequential.

Of course, our strike planning numbers were reviewed and approved by higher authority starting at the squadron level and working their way all the way to the Joint Chiefs of Staff in the Pentagon. This was a big, all-out, ass-kicking war which would probably play itself out in less than 24 hours . . . for better or for worse.

Finally, the thing occurred which I dreaded most. Someone in higher authority decided to try a few drills to see if we could really carry out our assignments. The drills started with loading out all the airplanes with weapons, which we did. Then the question was asked whether we could really get our ordnance into the air at the ungodly hours assigned. Our target had to be hit exactly five minutes after sunrise. This took best advantage of surprise, sun angle for the gunners on the ground and for the aviators trying to visually acquire the target. The first wave of this master plan had us all on dawn-plus-five-minutes target times.

The orders came in to launch with a full set of weapons, live and armed, at about four o'clock in the morning, to fly to a pre-determined spot 300 miles south as though we were really going to Cuba, turn around, divert to training targets for expending of the bombs, and finally to return to Cecil. As might be expected, we began briefing the mission at about 0100 and manned up at 0300 for the big launch. We all knew the added hazards of groping around in the dark over southern Florida with several hundred other airplanes, some coming from as far away as Texas and all with live ordnance.

A few of us even suspected that at the very last minute, just as we were rolling down the runway, someone would let us know that it was the real thing, that we were to really go to Cuba! So, it was an exciting night to say the least.

Our cell of six airplanes started engines and taxied out to the arming area in the dark where our ordnance crews were waiting to pull the safety pins and arming devices just before taking off. It was an eerie scene. The whole field seemed alive with activity. There were literally dozens of airplanes from different squadrons all starting up, taxiing, arming and moving into position for takeoff, all with our external lights on and flashing. I recall being very anxious about somehow not screwing up my switches, or the flight profile, or the target times or a hundred other things.

The launch sequence was a real war system which we had never tried before. To accommodate the large numbers of aircraft involved, we decided to taxi onto the runway with all six airplanes. The A-4s would line up in a balanced formation and make a division takeoff. The two F-8s would also line up, one of us on either flank of the A-4 division. We would execute a formation takeoff with each other ten seconds after the A-4s released their brakes. Overtaking the slower A-4s on takeoff roll, we would essentially be all joined up about the time the A-4s had raised their wheels and flaps a mile or so north of the end of the runway. This technique, not previously tried, would get the most airplanes into the air in the least amount of time. It was, I suspected, going to be dicey. I made myself concentrate on being extremely careful and watchful to be sure I didn't accidentally become a casualty of this poor man's war.

Finally, when we were next to go after the flight which had just taxied onto the runway, I made one last careful check of all my switches and procedures and decided that I was really ready to go to war. The A-4 division in the flight ahead of us rolled on visual signal and, 10 seconds later, the F-8 leader, Lieutenant Al Wattay, released his brakes. I noted with satisfaction that his wingman, Lieutenant John Nichols, was alert and watching, because he rolled a split-second afterward. I was excited and uncomfortable at the same time as I watched them pick up speed. They were already beginning to overrun the lights of the A-4s ahead of them. All of my senses were tuned up to the highest alert I have ever known. We were next to go!

Then it happened!

There is a weight-on-wheels microswitch on all Navy airplanes which overrides all armament switches as long as there is enough weight on the wheels to keep it closed. As the airplane breaks ground and the weight on the main mounts is slowly released, the weight-on-wheels circuit is opened and the position of all armament switches in the cockpit becomes operative.

The night was as black as the ace of spades. The overcast above us obscured any help which might have come from the night sky, moon, planets and stars. The two Crusaders were just beginning to break ground at about 150 knots. The whole world was in violent motion and breathtaking acceleration. In my peripheral vision I was watching the flight ahead of us. Suddenly the world turned from dark night to high noon as an enormous burst of white flame surrounded Al Wattay's airplane.

My night adaptation, such as it was, was gone in a nanosecond. I was almost totally blinded. The blinding source of light began to move rapidly forward in my peripheral vision and it became apparent as the light broke into two separate sources that the two giant roman candles were, in fact, two Sidewinders which had just come off the rails of Al's airplane.

In a state of semi-shock, I only had time to inhale before the two errant missiles roared by the four A-4s and then snuffed themselves out in the darkness ahead. Fortunately, the arming safety device in the Sidewinder requires that it fly about

3,000 feet straight ahead before the warhead can arm and the seeker head can become uncaged. That fact saved two of the A-4s because they were closer than 3,000 feet by then. Obviously, Al had done something to his cockpit switches other than what we had briefed. The weight-off-wheels switch had done its job.

From out of the darkness came a plaintive radio call, obviously from one of the A-4 pilots who had just watched the two missiles roar by his cockpit.

"Are we there already?"

Normally he was garrulous as a back seater. Tonight he was quieter. I suspect he felt the same uneasiness that I felt.

The stars were much clearer at our new flight level. But we were rapidly approaching the line of storm cells and I took the necessary precautions. I lowered my seat and turned up the intensity of the instrument panel and console lights almost to the maximum. Joe did the same in case he had to take over for any reason. We were reasonably prudent aviators, mainly because we appreciated just how rusty we really were.

The turbulence and lightning flashes began simultaneously, and the airplane began bouncing around pretty heavily. I recall a gnawing sense of uneasiness as though waiting for the first bad thing to happen. Would the violent turbulence cause a flame out? If it did, what would I do? I reviewed the emergency procedures to be sure I didn't miss something important.

It was deathly quiet as I fought with the controls, trying to keep the attitude gyro on an even keel as we bounced about like a floating cork in a storm. Joe was especially quiet. He had even turned off the hot microphone so I wasn't even listening to his breathing. All of my senses were turned up to full gain. I was waiting for trouble. Then, it struck, and my heart almost stopped beating!

The cold, clammy, unmistakable pincers of a huge lobster claw grabbed violently around the back of my scrawny neck, just above the flight suit collar and below the bottom edge of the flight helmet, right where the bare skin of the neck was exposed!

My reaction was purely atavistic. My survival instinct poured adrenaline into my veins and I lunged forward, letting go of the flight controls and grabbing at the claw with both hands. I wondered as I did so, how in the world did that son of a bitch get loose, and had he killed Joe in the process? Was that why I hadn't heard anything from poor old Joe?

Of course, all of this happened in less than a second, before reason could take over. It was then that I looked up into the rear view mirror and saw the silhouette of Joe's helmet and arms reaching around the back of my ejection seat and thrusting the lobster claw against the back of my neck. I flipped the hot microphone on and could hear Joe's hysterical laughter.

He had completely unstrapped himself from his ejection seat and stood up, leaning over the top of his instrument panel to get the lobster onto me. I almost wet my pants. I decided sheepishly that he had had his fun. Now, it was my turn!

He sat back in his seat still in gales of laughter, and I presume was starting to reconnect himself into the ejection seat. However, I struck first! I dumped the control stick forward violently and, looking in the rear view mirror, saw him flattened against the top of the canopy as the T-1A dove down into a storm cell. Then I reefed back hard on the control stick and heard the noisy clunk of his body slamming into the seat. He let out a guttural groan as he hit, and I knew it smarted. I repeated the maneuver three times to teach him a good lesson. On the third maneuver Joe let out a loud, "Holy Christ, Paul!"

I had to zoom the airplane at full power upward several thousand feet up to our assigned altitude. Air traffic had been light, I rationalized, and there was little chance of anyone else anywhere near us. Our IFF transponder did not have altitude read-out in those days, so our controller probably would not know what we had done. Nevertheless, he must have suspected something, perhaps a change in our radar return, because he called.

"Navy Jet Two Zero Seven, this is Washington Center. What is your altitude, over?"

"Flight Level four three five and climbing," I lied (we were a good thousand feet below that at the moment). I added by way of apology. "We hit a bad bump, Center, but we're okay now. We'll be back at Four Four Zero in 10 seconds, over."

"Roger, Two Zero Seven, how about a PIREP (pilot report) on the scrub, over." They were really nice and understanding about the deviation.

"Center, this is Two Zero Seven. Moderate to heavy for a moment there. Sorry, it got away from us for a moment. We are level Four Four Zero, Over."

"Roger, Two Zero Seven. No problem. Washington Center Out."

Meanwhile, back in the rear cockpit, Joe had lost his grip on the lobster and the wooden peg (which he had removed for his joke) and was searching for both with his flashlight. The poor creature was probably suffering from anoxia anyway. But Joe, a little bruised, had no way of knowing that. He found the lobster crawling around at the very end of the left rudder well. It took a great deal of delicate poking and probing with his pencil to retrieve the monster, put a piece of his pencil in the claw and get then thing back in its box. All through the retrieval process I could hear Joe grunting and muttering on the hot mike.

"Come here, you ugly son of a bitch," he had said more than once while coaxing it forward. I offered absolutely no help. And Joe knew better than to ask.

With our monster safely back in his cage and Joe nursing a few well-deserved bruises, we began our long descent to Andrews.

Several days later the Salins and the Gillcrists enjoyed a succulent lobster thermidor with plenty of good wine and a great deal of laughter and arguing over who got the best of whom at 44,000 feet over Wilmington, Delaware – Joe, the lobster, or me!

To complete the realism of the scene the pilot who made the series of carrier passes did them at the speeds and altitudes prescribed by our technical advisor, a former Japanese Zero pilot from the Akagi. Our tall Irishman landing signal officer, Jim Shannon, stood out on the LSO platform, paddles in hand, and performed the paddle-waving ritual upon which many of us had been weaned as fledgling carrier pilots. Of course, the use of paddles had long since been superseded by the optical landing system. So it was with a certain wistful nostalgia that many of us watched the old-time carrier approaches and Jim Shannon's paddle waving.

It was, in a sense, a re-enactment of a scene in real life that most of us never thought we would ever see again. As I watched Shannon's lanky figure, his flight suit rippling in the 30-knot gale giving the time-honored paddles signals, I could see that he was even putting into the act all of the body English that LSOs had used.

Each LSO, waving paddles in those days, had his own personality and signature that no other LSO would dare to emulate. It was an art form of visual communication the likes of which we will never see again.

That night we turned in early for the second and last launch sequence. The director informed me that he had gotten good film coverage of my deck launch and settle off the bow sequence, but he wanted one just as good the next time around. I told him I was going to modify my deck launch procedures a little bit, strongly recommending that he take extra good care of yesterday's film. He got my meaning!

As we prepared for the second and final shooting of the launch sequence, a minor maintenance problem surfaced. I was sitting in Ready Room Three reading when one of our maintenance chiefs came in with a pained look on his face as he approached Billy Phillips.

"Captain" he said, "We've been losing propeller spinners because the fasteners that hold them on are backing off. What we need are lock nuts. I've found where there are some in the Supply Department but the Chief in charge down there won't issue them to me because we don't have a unit identification code (UIC).

"Of course, we don't have a UIC, Chief," responded Billy. "20th Century Fox ain't in the God-damned Navy. But that's no reason for not giving us a few lousy nuts. It's not like we're asking for the God-damned Taj Mahal."

"Captain, I explained all that to him but he won't listen."

"Okay, Chief, I'll handle it." said Billy, reaching for the telephone on the duty officer's desk.

"Hello, Chief," Billy began in his most pleasant voice. "This is Captain Phillips, United States Navy, Active Duty. How are you?" There was a short pause. "Listen, Chief, I understand you refused to give Chief Johnson a few 5/16 lock nuts a few minutes ago. Is that right?"

*Author (left) skipper of "Zero" squadron, and Ed Klapka (center) skipper of "Kate" squadron, plus three other participants at the world premier of "Tora, Tora, Tora" in Washington, D.C., December 1970.*

The last question was followed by a long pause. "I understand all that UIC stuff, Chief, but you know that this movie crew is a commercial outfit and couldn't possibly have a UIC. But, they need a little help and we ought to give it to them, okay?"

Another pause ensued. Billy's face began to redden as he listened. Finally, he said, "Listen very carefully, Chief. I'm coming down there right now and you are going to give me a handful of those nuts or I am going to have a handful of your nuts. Is that clear?" With that, Billy slammed the receiver down and stormed out the ready room door. Five minutes later he re-entered and, walking over to Chief Johnson, dropped a handful of the lock nuts into his hand. We all turned into our bunks early that night in anticipation of a 3:00 a.m. wake-up.

"Alpha Two One Three, this is Steamboat Tower, how many dummies do you have on board?" I smiled at the radio transmission because of the exasperated tone of the air boss' voice. None of YORKTOWN's personnel had any patience with us. An uncooperative attitude seemed to pervade the whole ship. It was obvious that the ship's leadership thought we were one giant pain in the ass. It was too bad . . . a mark of poor leadership. Given that the decision to use YORKTOWN had already been made by the Chief of Naval Operations, all of the griping in the world wasn't going to change one damn thing. Here was an opportunity for the ship's commanding officer to turn it into a fun event for the crew. But in the six days that I was on board, I never saw the ship's Captain, or, for that matter, dealt with any of his officers. I found that unusual, since there were half a dozen active-duty Navy Captains and at least a dozen Commanders.

But this morning's launch was going to be different from Tuesday's. For one thing, I told the plane captain that if I even saw a flash camera anywhere on the flight deck, I was going to stomp it into smithereens.

The events leading up to the starting of engines and the rolling of cameras were identical to events the previous Tuesday morning. My engine was just warming up when I heard the air boss' unusual question. I also knew that he was calling Ed Klapka, the commanding officer of the Kate squadron. Each of the Kates carried two anthropomorphic dummies in their back seats simulating Japanese air crewmen. In Ed's airplane, one of the dummies had been removed to make room for the Japanese flight coordinator in yesterday's carrier landing sequence. I knew that the dummy should have been put back in. Someone on the movie team must have questioned that fact. Now the air boss was trying impatiently to find out.

Ed Klapka, whose airplane was parked a few feet from mine, was busy trying to start his engine and, I suspect, hadn't heard the air boss' question.

"Two One Three, this is Tower, how do you read me?" The air boss' voice really sounded irritated this time.

"Tower, this is Two One Three, I read you loud and clear. How me, over?" Ed had apparently just gotten his generator and radio on the line.

"Roger, Two One Three, this is Tower loud and clear. How many dummies do you have on board, over?" There was a pause, and I'm sure Ed, like me, was wondering what he was doing sitting in a 50-year old clunker, on the flight deck of an aircraft carrier at four o'clock in the morning, about to risk his life for a few lousy bucks. His response nearly broke me up.

"Steamboat Tower, this is Alpha November Two One Three. The only dummy in this airplane is me. Out"

The launch sequence went as smooth as glass. My settle off the bow maneuver was a little more conservative, and I'm sure the film coverage of it ended up on the cutting room floor. At the world premiere in Washington, D.C., I noticed that the director had selected the sequence from Tuesday's shooting.

After landing at North Island we completed the long taxi back to the carrier pier for the third, and last, time. I climbed out of the cockpit feeling no small amount of regret. Most of the pilots were going to fly out to Hawaii to participate in the more exciting strike scenes. I wanted to go in the worst way, but couldn't. My leave was up tomorrow and I had to catch a commercial flight back to Washington, D.C., that evening. The honeymoon was over, at least for me. The four pilots in my division were posing in front of my airplane when my wingman, Guy Strong, came up with a great idea.

"How about a cool one?" he asked.

Without another thought, we all piled into one of the rented sports cars and headed for the main gate. Never mind that it was only 8:30 in the morning. Never mind that we were wearing Japanese flight suits, make-up, and fur-lined cloth flying helmets with goggles. This was the end of my movie career, we were all thirsty and it somehow seemed like the right thing to do! We cruised down Fourth Street to the main drag and turned left, down where the old nickel snatcher used to be and pulled up in front of a flea-bitten bar called Mitzi's Place. It was just beyond the Mexican Village.

I expected to cause a stir when we walked in in our movie outfits. Not so! There were two characters slumped over bar stools having their morning constitutional beers. We ambled over to the bar and ordered four beers. The poker-faced bartender served us without the slightest show of interest. It tasted good. Better than I can remember in a long time. Guy Strong raised his glass in a toast, knowing that he was going on to Hawaii and I was not.

"Here's to the best deal any carrier aviators ever had."

"Hear, hear," the other two said.

"Amen," was my response.

As a postscript, Guy Strong was killed a few weeks later when the Zeke he was flying crashed during a filming of the strike sequence. He was a hell of a fine young man!

prayer that I wouldn't "screw it up," and asked Pat to forgive my irreverent thoughts.

Then I remembered my embarrassment the night before at the wake. Rita had asked me to lead the group in a recital of the Rosary. "Of course," was my reply. The family rosary had been a regular event in my family as I grew up. Though I never said so, I always thought it was kind of silly to say the same prayer several hundred times. I could never force myself to concentrate on what I was saying after the first few "Hail Marys."

Nevertheless, just about every evening after supper someone would say in a loud voice, "Rosary." I think the culprit was my sister, Mary. Then all eleven of us would troop into the living room, kneel down and recite all 15 decades. One hundred fifty Hail Marys in all, to say nothing of Our Fathers and Glory Be's.

So, I came to memorize the 15 mysteries of the Rosary; five joyful mysteries, five sorrowful and five glorious. Last night, with the whole Tacke clan gathered in a room with Pat laid out in the casket, I knew I was going to forget at least one of the mysteries . . . and I did. I got through the joyful mysteries okay, but I stumbled when I got to the third sorrowful mystery.

Pat's father, kneeling next to me, must have anticipated my impending loss of memory because, after only a fraction of a second's hesitation, he whispered just loud enough for me to hear, "The Crowning of Thorns." I picked up on it so quickly that I thought I may have gotten away with it. I raised my eyes while reciting. Our gazes met and I caught an ever so slight twinkle in his eyes. The silent communication in that brief glance said, "Young man. You haven't said a Rosary in years."

The requiem mass was said by Father Verhoeven, pastor of Saint Mary's Roman Catholic Church. His ministry extended beyond the confines of Cottonwood to several of the other surrounding small communities on Camas Prairie. A large percentage of its inhabitants were devout Roman Catholics of German descent who had emigrated to the verdant Camas prairie from the farmlands of the Midwest in the 1880s and 1890s.

"Pat would be pleased," I thought as I watched the casket move down the aisle toward the altar, flanked by his four brothers and two of his friends. The ceremony was brief, the homily was simple and the readings appropriate for the occasion. As expected, most of the congregation received Holy Communion.

After the service the pall bearers delivered the casket to Cletus' hearse and I got in behind the wheel of the flower car.

As Cletus predicted, the whole town showed up for the interment. The procession from the church to the gravesite went off without a hitch. Every family car in Cottonwood fell in behind the hearse with their lights on. The slow procession to the small cemetery brought Cottonwood's first casualty of the Vietnam War to his final resting place. The gravesite was on an elevation which overlooked the town of Cottonwood and provided a panoramic view of Camas Prairie and its breathtaking backdrop of the Bitterroot Mountains to the east. The dome of the deep blue, cloudless sky was Pat Tacke's basilica. The bereaved family sat in a row of folding metal chairs beside the gravesite. A rectangular patch of ground had been cleared of snow for the several rows of chairs. Rita placed Pat's Air Force cap on top of the casket. The wind had died down so that the priest's rendering of the 23rd Psalm could be clearly heard by the entire congregation.

"How many times have I done this?" I asked myself. "How many times in the last three years have I listened to 'The Lord is my shepherd. I shall not want . . .' intoned over the remains of a friend? How many times have I wondered who will be standing there when it's me beneath the draped American flag'?"

Here I was listening to it again for a friend. I noticed two old Gonzaga class-mates in the crowd who must have driven all the way, one from Portland, Oregon, and the other from Spokane, Washington. After all these years I wondered how in the world they ever found out about the ceremony.

The Air Force honor guard did a fine job. As always, the bereaved sitting in the front row of folding chairs, flinched involuntarily when the first volley of shots rang out. The women's shoulders began quietly quivering when Taps was played. During Taps I looked around at an entire community of men of the soil, with cal-loused hands and weather-beaten faces, standing in their Sunday-go-to-meeting suits and good oxfords in six inches of snow.

No one moved. The stillness between bugle notes was overwhelming. The last notes of Taps trailed away into the silence and the Honor Guard began the meticulous procedure of taking the flag from atop the casket and folding it into a tight triangle. The head of the Honor Guard, holding it stiffly in front of him, walked ceremoniously over to me and gave it to me. When I took it from him he gave me a snappy salute. I returned the salute, then carried the flag over to where Rita was seated, stood in front of her and said, "Rita, on behalf of a grateful nation, I present this American flag to you." Rita took the flag, I saluted her and then walked back to where I had been standing with a terrible empty feeling in the pit of my stomach. All I could think of was an irreverent curse and the thought that "This Goddamned war has finally touched this beautiful place and these wonderful people!"

Edwin Markham's famous words came to me in a rush:

*"Here was a man to hold against the world,*
*A man to match the mountains and the sea.*
*As when a lordly cedar, green with boughs,*
*Goes down with a great shout upon the hills,*
*And leaves a lonesome place against the sky."*

# 26

## CLEAR THE PATTERN

I t was about nine o'clock on a bright spring morning when we taxied out of the line with Charlie Pitman in the front seat and me, talking to Andrews Ground Control, in the rear.

"Andrews Ground Control, This is Marine Jet 29616, ready for takeoff. Over." We had received a rather lengthy flight clearance to Nellis Air Force Base near Las Vegas, Nevada, with a refueling stop at Tinker Air Force Base near Oklahoma City. Andrews ground control told us to switch to tower frequency for takeoff.

Meanwhile, Charlie had been running through the takeoff check list, reading the items out loud over the intercom. Some of the items I was supposed to respond to, like ejection seat pins, for example. We turned the corner and entered the warm-up area for the north runway just as Charlie told me the check-off list was complete and we were ready to "go flying."

I switched to tower frequency, got clearance onto the duty runway and was told to hold my position. A moment later the tower assigned us a departure control frequency, gave us our departure instructions and told us we were cleared for take-off. Charlie, waiting for this, had already run the power up. He released the brakes and we started trundling down the runway. We were, indeed, going flying. As we lifted off I switched frequencies and checked in with departure control while Charlie adhered to our departure instructions.

"Andrews Departure, this is Marine Jet Two Nine Six One Six, passing five hundred feet for two thousand five hundred. Over." This was our initial radio call as we checked into the system. They responded, acknowledging that they had us in radar contact.

Moments later, when we had followed the radar vectors assigned and were approaching 2,500 feet, I called and told them so. They switched us to an air route traffic control (ARTC) frequency and cleared us further up to 5,000 feet.

I checked in next with Washington Air Route Traffic Control Center (ARTCC). A nice, crisp voice acknowledged my call, told us he had us in radar contact and cleared us all the way to Flight Level Three Nine Zero (39,000 feet).

The complicated process of getting out of the busy Washington, D.C., area under positive radar control was almost completed. But it took all of my attention on the radio, and Charlie's piloting skills and adherence to departure instructions to accomplish the job.

It promised to be a pretty day and I recall feeling good about our trip. As the saying goes in the Pentagon, "Anytime spent outside the beltway counts as leave." Our trip to Nevada promised to be pleasant because we would get a fair amount of flying, it would be interesting and, of course, it was "outside the Beltway." What could be better?, I asked myself. Charlie, I'm sure, felt the same way I did. On the next leg, the one from Oklahoma City, I would get to fly in the front and he would do all the radio and navigation stuff from the back seat. That was our arrangement for our "joint" programs. It was a Marine jet we had "borrowed." Therefore, in accordance with Marine Corps protocol, Charlie was always in the front seat when we left and when we returned to Andrews. After that we alternated. It was a good arrangement from both our viewpoints.

I was just beginning to relax a bit when the explosion occurred! The explosion was followed by a very strong engine vibration. Over the intercom I heard Charlie's voice telling me something I couldn't understand because of the noise . . . but I distinctly heard the words, "flame-out!" Only seconds had elapsed since the explosion, and my eyes swung to two important engine instruments; the rpm (revolutions per minute) gauge and the turbine inlet temperature gauge (TIT). They were both unwinding steadily . . . a sure indication, I decided, of flame-out. Without further ado I keyed the radio transmitter button.

"Washington Center, this is Marine Jet 616 passing fourteen thousand; our engine just flamed out. We are commencing a descent and declaring an emergency. We are overhead Dulles and have the field in sight. We intend to make a dead-stick landing at Dulles. Request you clear the pattern. Over!"

"Jesus Christ! What the hell did you say that for?" Charlie asked me incredulously on the intercom. We had somehow leveled at 9,000 feet and the engine vibration had diminished a little . . . and the engine seemed to be running. I responded, more than just a little bit nettled.

"God damn it, Charlie. I heard you say 'flame-out.'"

"What I said," Charlie responded, sarcasm dripping from each word, "Was, I think we may be about to flame-out. Can't you see the gauges, for Christ sake?"

# PART VI

# THE IRON ANGELS

*"... and though you be done to death, what then?*
*If you battled the best you could;*
*If you played your part in the world of men,*
*Why, the Critic will call it good.*
*Death comes with a crawl, or comes with a pounce,*
*And whether he's slow or spry,*
*It isn't the fact that you're dead that counts,*
*But only, how did you die?"*

*– Edmund Vance Cooke*

*Fighter Squadron FIFTY-THREE (VF-53), the Iron Angels, had a history which went back to World War II and the Pacific Campaign. It was one of the most distinguished fighter squadrons in the United States Navy. Of course, every fighter squadron worth its salt will, and ought to, make that sort of claim! After all, that is why they are out there on the cutting edge of United States' foreign policy.*

*Nevertheless, the squadron had a reputation for being a particularly well-seasoned combat organization. The average number of Crusader flight hours boasted by pilots in the squadron was considerably higher than for a normal fleet squadron.*

*Often misunderstood by the general public (the taxpayer) and the Congress is why we have armed forces in the first place. We have them, quite bluntly, to kill people and to destroy things. If they can't do that well, then nothing else that they do is really very important.*

*The Vatican has this group of bumpkins who go about caparisoned in medieval garments carrying medieval weapons. Their original purpose was to protect the Pope. Nowadays, they couldn't even enforce parking regulations in Saint Peter's Square. They are what is called, in current military parlance, a symbolic force!*

*The British Royal Navy, once feared around the globe for its power and competence, is now also a symbolic force.*

*The current Presidential administration, lured by the temptations of "collective security," is in the process of turning the United States Navy into another "symbolic force." That, if it happens, will truly be the saddest legacy of the 1990s!*

---

This part of the story contains a few nuggets showing what it was like to be a part of a squadron full of professionals who understood that they were supposed to be good at punishing the enemy! They went about the business of conducting combat operations in Southeast Asia with a degree of professionalism which I had never found in another fleet squadron. They accepted their daily target assignments over North and South Vietnam as the standard fare of fleet operations in a combat zone. There weren't any malingerers or shirkers. There weren't any pilots who would not go over the beach. There weren't any who demonstrated a propensity for "downing" their airplanes whenever they were assigned to a difficult or hazardous target.

They just accepted their share of the easy ones along with their share of the hard ones. This was because they were professionals who knew that if they "tanked" a tough mission, one of their squadron shipmates would have to take it. I was proud of every one of them. And, when we returned to the states after my tour as C.O. I was extremely proud to note that they all came home . . . alive!

The "Iron Angels." Author is sixth from right, rear row – Tonkin Gulf, September 1968. (Author's collection)

Alpha Strikes in Route Package Six required sufficient ceilings to remain above the critical minimum safe altitude for small arms and light automatic anti-aircraft-artillery (AAA) of 3,000 feet above the terrain. The visibility under these ceilings had to be great enough to visually acquire surface-to-air missiles (SAMs) in enough time for evasive maneuvers.

From the beginning of December to the beginning of May, the Tonkin Gulf northeast monsoons brought torrential rains. Poor visibility and cloud cover could extend from 500 all the way to 30,000 feet.

During May, the wind shift to the southwest brought generally clear skies. The movement of supplies from North Vietnam to the south increased dramatically during the northeast monsoons when air interdiction was restricted by weather.

As Rolling Thunder operations moved steadily northward and North Vietnamese air defenses increased in quantity as well as quality, the composition and tactical development of Alpha Strikes evolved. Certain elements of strike groups were assigned specific missions. Two-plane sections of attack aircraft operated against the radar emitters of SAM and AAA sites. These sections, called Iron Hand, were equipped with Shrike AGM-45 anti-radiation missiles which homed on the antenna source of radar emissions. The Iron Hand shooter needed an escort with a second set of eyes, since the Shrike launch parameters were achieved by monitoring instruments inside he cockpit, and often times the launch aircraft and the SAM site would be literally jousting with one another. The time for executing a Shrike launch maneuver, firing the Shrike then dodging the SAM was often reduced to a few short seconds.

Another specified mission was flak suppression using forward-firing rockets or anti-personnel cluster bombs to silence AAA in advance of strike aircraft. Fighter escort and MiG sweeps were employed to protect strike groups from hostile aircraft, and sophisticated stand-off electronic jamming aircraft were used to render North Vietnamese acquisition and fire control radars temporarily ineffective.

In the spring of 1966, when the Air Wing FIVE operations officer briefed the concept and execution plans for Rolling Thunder to the air wing pilots, the code name took on new meaning. One wag in the audience, intimidated by the booming voice, fierce visage and massive frame of the briefer, dubbed him "Rolling Thunder." For the rest of his Navy career, Jack Perkins kept the nickname.

Lieutenant Commander John Cooley Perkins was one of those unforgettable characters. I first met him in the spring of 1950 when the new second-year midshipmen were allowed to try out for the Naval Academy varsity football team. The spring practice was unusually arduous because of the terrible combination of temperature and humidity. I was immediately taken by Jack. First impressions are often lasting ones!

Jack was standing with all the new second-year men (youngsters) listening to the coach give his standard opening-day lecture. I, a third-year man and a member of last year's varsity squad, was giving the new guys the "once over." Jack stood there listening to the coach, apparently unaware that they were being scrutinized by the upper classmen. He had his hands on his hips and was clad in the standard gray workout shorts, white tee shirt stencilled "stolen from the Naval Academy Athletic Association" and football shoes. His close-trimmed head blended into his powerful shoulders by way of a 17-inch neck. Jack's barrel chest and powerful shoulders made him look like 30 pounds of clams in a 20-pound sack. His thighs looked like the Cedars of Lebanon. He had the overall appearance of a six foot fire hydrant with a feral expression on his face. I thought to myself, this guy looks like he really means business.

Jack made the varsity squad and went on to become one of the meanest, toughest and hardest-hitting linebackers in Naval Academy history! He took great delight in manhandling blockers and tackling ball carriers so hard they usually took extra time getting to their feet and limping back to the huddle while shaking their heads. Jack's booming voice could be heard all the way up in the stands as he urged his teammates to greater efforts. He really enjoyed "kicking ass and taking names" as he so often put it. Jack must have had a terribly intimidating effect on opposing quarterbacks.

After my graduation, our career paths separated and it was 13 years before Jack and I met again as the air wing prepared to sail again, this time in USS HANCOCK on another combat deployment to southeast Asia. Jack, now the air wing operations officer, was responsible for all air wing strike planning and for the execution of the ship's daily air plan by the wing squadrons. Jack carried out his duties with the same vigor that he had displayed on the gridiron years earlier.

But the intervening years and Jack's zest for life had taken their toll. On more than one occasion, I heard the air wing commander mention to Jack that he was carrying a little too much weight and ought to "knock off a hundred pounds or so." Jack was now a mountain of a man and even more intimidating than ever. To see him sitting strapped into that tiny A-4 cockpit made me think of a cork in a bottle. Jack manhandled that little Skyhawk the same way he used to stack up blockers . . . with authority!

Jack loved to go on liberty. It was always apparent when he was on the town. After a couple of beers, his booming laugh could be heard a quarter of a mile away. He had a wonderful sense of humor and was fun to be around. However, the air wing staff was a small group, and Jack often went ashore on liberty with the air wing landing signal officer, Lieutenant Commander Dave Albritton.

Dave was the antithesis of Jack, all of the things Jack was not. He was short, slim, quiet and reserved. To see the two of them leave the quarterdeck together and go down the accommodation ladder was an amusing sight.

# 28

## MiGS OVER BULLSEYE

F ar and away the most boring, tedious and uninteresting fighter mission in the Tonkin Gulf was the barrier combat air patrol (BARCAP) mission. But, every so often they became intensely exciting. The tactical commander in the Tonkin Gulf maintained a BARCAP of two fighters on station just off the port of Haiphong anytime there were carrier battle group forces on Yankee Station. The purpose of the BARCAP was to provide immediate response to any threat to those forces of any kind.

The routine was simple. The BARCAP would be launched from the Yankee Station carrier, proceed north to relieve the off-going BARCAP and orbit off the coast of Haiphong clear of the SAM/AAA envelope. About halfway through a typical BARCAP mission, a carrier-based aerial tanker would rendezvous with the fighters and tank them with a couple thousand pounds of fuel to keep them combat ready until they were relieved by fighters on the next cycle. In those days, the carriers were operating a one and a half hour cycle time.

Since the fighters were launched first and recovered last, a typical BARCAP flight would last one hour and forty-five minutes, and the only interesting parts were the catapult shot, the aerial tanking and the carrier landing. However, once in a blue moon a BARCAP got to be interesting, and when it did, it got very interesting.

On 22 June 1967, my wingman, Lieutenant Rick Harris, and I were launched on a routine BARCAP mission. No sooner had we relieved the off-going section of Phantoms from the other carrier than a call came in from Red Crown, the radar-equipped cruiser telling us that enemy aircraft from the North Vietnamese airfield at Kep were proceeding on a northeasterly vector directly towards us. Red Crown, the radar equipped cruiser, directed us toward the contact.

I pushed the throttle all the way to the firewall and headed toward the mouth of Haiphong harbor. Calling up my wingman, I said; "Two, cross over to the right side, and 'knockers up." I wanted my wingman down-sun so I could make the best use of his lookout into afternoon sun, thus enabling me to concentrate on my radar scope. Making early radar contact on this guy was going to be the key to killing him. The "knockers up" call was to remind my wingman to turn on all of his fire control switches.

We crossed the beach at 18,000 feet and 450 knots. There would be no "feet dry" call to Red Crown because I didn't want the North Vietnamese who were listening to know about my movements. My heart was beating a mile a minute. Hot damn, I thought, this may be my day!

Red Crown told us that the enemy aircraft was "twelve o'clock at 30 miles" and approximately the same altitude, cruising at 520 knots. I did a quick mental calculation. At a closure rate of almost 1,000 knots, the MiG would be in visual range in about 90 seconds.

"Damn," I swore to myself. "Still no radar contact. Where the hell is he?"

At this point, a little voice in the back of my skull began sending me warning signals. Why hadn't I heard any enemy fire control signals on my radar homing and warning systems? I was already 10 miles inland, in the heart of the envelope of at least six different surface-to-air missile sites . . . and no tones of any kind were heard in my headset. Another concern was the big thunder cloud which we were about to enter. Once in the cloud our problems would be compounded. By now, my mind was racing a mile a minute.

Dodging SAMs is difficult enough when you can see them. However, when you are in the clouds, it is quite another story. At that juncture Red Crown called me to say that the MiG had started a right turn. That did it! The North Vietnamese had used MiGs to lure Navy fighters into SAM traps before. Billy Phillips had chased a MiG over downtown Hanoi and all he had gotten for his trouble was half a dozen SAMs fired at him, several of which came close to bagging him.

I rolled my F-8 into a 90-degree bank and pulled a six g reversal turn to the right. No sooner had I crossed the beach outbound than the MiG again reversed course and the sequence repeated itself. The second time outbound, I decided to fool the enemy radar. As we crossed the beach outbound, without any radio transmissions, I rolled the F-8 inverted and did a "split-s" maneuver, pulling out at about 10,000 feet and 600 knots, heading back towards Hanoi, my head buried in my radar scope. I knew that the enemy radar controllers, watching my radar return, would see it appear to stop, then I would go in the other direction. They would have less time to warn the MiG to turn away.

"I'll get that little son of a bitch this time," I muttered into my mask as I played with my radar antenna tilt, looking up to where I hoped he would be. No luck! Still no contact!

Suddenly, as the thunder cloud loomed closer, I heard an acquisition radar on my headset. Looking off my right wing, I saw Rick Harris' airplane and wondered what he must be thinking. Here we were at the outskirts of Hanoi at 10,000 feet, doing 600 knots and about to punch right into the middle of a huge "thunder bumper."

I dipped my left wing once, a signal to Rick that I was about to turn left, then rolled into a steep turn in that direction. I had decided to quit screwing around. Any moment now, I expected a SAM launch signal, and the hair was tingling on the back of my neck as I turned our tail to the threat and crossed the beach outbound for the third time.

"Red Crown, this is Firefighter Two Zero Two. Recommend we knock this off. Over." My concern for being SAM bait was apparently shared by him because he agreed.

"Firefighter Two Zero Two. This is Red Crown. Concur. Return to CAP station. Out."

The return flight to Rampage (U.S.S. HANCOCK) was uneventful. As I sat in the ready room reading the message board after debriefing Rick, I could hear him in the back of the ready room animatedly describing having areed around Hanoi with the skipper on a "routine" BARCAP mission!

# 29

## A BARCAP WITH THE SKIPPER

Everyone got an even share of the BARCAP missions since they were so onerous. The very last day of this line period was fairly quiet and the last cycle of the day found me scheduled for a BARCAP mission with Lieutenant Commander Bob Rice, the squadron safety officer. Lieutenant Commander Sandy Button, the squadron maintenance officer, was the spare pilot. I had briefed the standard BARCAP mission with the exception that I had requested that the ship's air operations officer launch the spare airplane. This was for the purpose of getting Sandy Button (a new arrival) a few extra carrier landings. The last launch of the day was not a bad time to occasionally make such a request.

This particular BARCAP mission was progressing uneventfully. A scattered cloud deck at about 10,000 feet had begun to thicken into a broken layer. I was able to locate myself visually through a hole in the cloud layer through which I could see the end of a string of tiny islands which marked the westernmost edge of the patrol station.

In order to liven up an otherwise boring flight, I had briefed that when we came to the end of each patrol leg we would do "in-place" turns. This was a semi-precision maneuver in which the two airplane elements simultaneously turn 30 degrees away from each other, lower the nose 10 degrees, tap a little afterburner to pick up 50 knots, then pull up and turn into each other in a nose-high wingover maneuver.

If the maneuver is flown with precision, the two elements cross paths a few hundred feet apart at the apex of the maneuver, and finish up having swapped sides of the formation and reversed course.

*USS ORISKANY (CVA-34) Tonkin Gulf, May 1967. (Official U.S. Navy photograph)*

Since my flight had the spare airplane along, Bob Rice flew as my wingman and Sandy Button flew the second section as a single airplane. Bob had decided to play "Blue Angel." Instead of flying the prescribed "free cruise" formation, he flew a "parade" (very close) formation. I didn't say anything when he tried it on the first "in-place" turn, so he repeated the performance in all subsequent turns. Although it was not a useful tactical formation, I let him play. It seemed relatively harmless.

We were at the western end of the patrol station, the one closest to the North Vietnamese coast, when I called for an "in-place" turn. We were at the very apex of the maneuver, nearly inverted at about 25,000 feet and as slow as 250 knots. Bob was enjoying a modicum of success staying in the correct parade formation on my right wing. I recall thinking, admiringly, that he was "pretty good."

Suddenly, to my utter astonishment, a SAM alert tone came on in my headset, followed immediately by a launch alert signal (meaning a missile has been launched). I couldn't believe anyone would shoot at us since we were several miles outside any known SAM envelope.

"Jesus Christ," I muttered in my mask. "This is the worst possible time for this!" Here we were, almost inverted at 25,000 feet, slow, and with my wingman playing Blue Angel. Both Bob and Sandy heard it too. A quick glance to my right showed that Bob had moved out (in less than two seconds) to a proper tactical formation for SAM evasion. I tapped burner to get us some airspeed quick.

"Two SAMs coming through the cloud deck at nine o'clock!" It was Sandy's excited voice. It was also the first SAM he had ever seen.

I looked over in that direction and, just as sure as hell, there they were! They were rising majestically, in parallel, vertical paths toward us and accelerating rapidly. There was absolutely no doubt about it! We were their target!

Our airplanes were picking up speed and I rolled to an upright attitude in a steep dive, setting up the proper geometry for a good defensive SAM evasion maneuver. The SAMs had now leveled off at our altitude and were traveling with startling speed directly at us. I was estimating that were two or three seconds from initiating the maneuver when both missiles exploded. Their range from us was probably three miles.

We assumed later that the SAM battery commander must have decided to waste those two "birds" at us, even though they must have known they were out of range, just to "get our attention!"

Our return to Rampage was uneventful. Bob Rice got to the ready room before I did and had written on the blackboard in large letters:

"NEVER FLY A BARCAP WITH THE SKIPPER!"

The launch and rendezvous were accomplished quickly and professionally, and the flight leader, assisted by vectors from the E-2, headed us in towards the beach. We cruised inbound at an altitude of 20,000 feet. At about 30 miles off the beach, the strike airplanes began a gradual descent at full power to help the under-powered A-4s pick up some maneuvering speed.

As briefed, George and I stayed at the original cruise altitude, but went to full military power and began a section "weave" over the descending strike force as our own speed increased. The strike group leveled at 8,000 feet and crossed the beach directly over the designated coast-in point. The power plant, railroad yard and flak sites were all clearly visible. I remember the strike leader's admonition to me to take careful note of their bomb hits since I "wouldn't have anything better to do" during the strike!

Shortly after the strike group crossed the beach they began jinking, and I noticed small black puffs bursting around them . . . obviously 57-mm AAA fire. About two minutes after coast-in, the strike leader called "rolling in" on his run. George and I, jinking ourselves now, had just begun a sweeping turn around the eastern side of Nam Dinh. So far it had been a "milk run" and I focused my attention on the power plant and flak sites so that I could report the bomb hits.

During a brief glance at George's airplane, I saw a huge black shell burst about 200 feet directly behind his airplane. That one was followed in rapid succession by three more bursts, each one about 50 feet closer to his tailpipe. The inexorable march of those menacing explosions up the wake of his airplane was so fast that I barely had time to push the microphone button and shout.

*F8E returning aboard the USS ORISKANY (CVA-34) Tonkin Gulf September 1966. (Official U.S. Navy photograph)*

"Two. Break left!" George, God bless his soul, had the reactions of Sugar Ray Robinson and broke his airplane directly into me almost as the last syllable escaped my lips . . . and it was a good thing he did so. A fraction of a second after he broke, the airspace which he had just vacated erupted in a burst of black smoke. Meanwhile, I had at the same time broken left, and had temporarily lost sight of him somewhere under the belly of my airplane.

"One. Break right!" Came George's breathless rejoinder . . . and I did so. It was then that I realized that, for some unfathomable reason, at least two of the 85-mm flak sites in the Nam Dinh area had decided to take George and me under fire.

The next two minutes or so were some of the longest minutes of my life. The air around our two airplanes was filled with very large, black explosions. We jinked left and right, up and down with both of us now in full afterburner to keep up our speed. No matter which way we turned, or what we did, the gunners seemed to generate an immediate and uncannily accurate tracking solution. We never got to roll out of our turning maneuvers . . . not even for a second. It was one long series of frantic break calls which, of course, were recorded by the E-2 and relayed back to the strike center on the ship.

It has been said that the passage of time is relative. The two seconds that one may spend sitting on a hot stove may seem like an eternity. . . but is only seconds to the person watching him. The two minutes we spent over Nam Dinh that beautiful, clear, crisp morning were the longest of my life.

I found myself halfway through the evolution wondering where those lumbering, snail-like A-4s were and, were they ever going to finish their attack and get out of this infernal target area? Time had truly slowed down for me and George. My God!, I wondered. How long did it take to make a lousy bombing run, anyway?

Finally, after what seemed like an eternity, I saw the last A-4 streaking toward the coast at tree-top level. The first "feet wet" calls were already being made when I turned in a descending, full-afterburner arc toward the inviting brown waters of the Tonkin Gulf. Somewhere during the melee I recall looking down just in time to see a large ball of fire erupt inside the walls of the large enclosure just behind the power plant's main building. I had not the slightest idea whose bomb drop it was, nor did I care!

Now, however, as we were egressing, I began to feel sheepish over how poorly I had observed the sequence of events of what had been a fairly simple, straightforward attack! I began to wonder what I was going to say when it came my time to speak at the general strike debrief.

I was just stepping over the "knee knocker" and into the small space identified as "strike operations" when the captain of the ship came out. He had apparently been in the space during the strike and listening in on the strike radio frequency relayed back to the ship.

on the pods. Thereafter, the pods, now with flat fronts, represented a huge drag count and had to be jettisoned as soon as possible. The problem was that the allowable jettisoning parameters were so restrictive that they really couldn't be jettisoned in the target area.

The only approved jettison parameter was 290 knots in straight and level flight. Since no Crusader pilot in his right mind would be caught dead flying at that speed over enemy territory, it meant lugging those high drag pods out over the water before slowing down and getting rid of them. The pods tended to do crazy things If we did not adhere strictly to the jettison parameters. I saw one pair of empty pods, when released, proceed forward, up and over the top of the Crusader's wing, bump along the wing's upper surface and then disappear from view, damaging the wing in the process.

My wingman, "Skip" Umstead, let two of them go one day while I was crossing behind him in maneuvering flight. I was startled to see them come off his airplane and appear to accelerate toward me. There was just time to stomp on one rudder to permit one pod to pass just beneath my fuselage. The other passed a few feet over the top of my canopy. To say that I was annoyed was an understatement! I said a few obscenities into my oxygen mask that day!

The four-pod LAU-7 configuration was also great for coastal reconnaissance missions. The wiring in the Crusader allowed firing the pods one at a time, with a few minor modifications by an electrician. This permitted options of firing in combinations of one through four pods at a single target.

On 11 March 1967, George Talken and I went on a coastal reconnaissance mission in Route Package Five, the area just south of Haiphong. It was a gloriously clear and beautiful day for flying, with recovery scheduled for just after sunset. I was concerned about fuel because of the high drag count of the expended rocket pods, so we established a "HOWGOZIT" chart for our kneeboards which displayed a graph showing time until recovery versus fuel remaining. If at any time on our reconnaissance mission we fell below the line, we terminated the mission and returned to the ship, jettisoning our rocket pods enroute home. This, of course, assumed that we failed to find suitable targets for our weapons load.

We flew to the northernmost end of Route Package Five and began reconnoitering south, keeping about 10 miles off the beach and flying at about 4,000 feet and 400 knots. As luck would have it, we didn't find anything for the first 45 minutes of the reconnaissance portion of the flight. We were just passing the mouth of the Hua Loc River near a place where two tributaries form an hour-glass shape, and was known as "the hourglass." The sun was just a few minutes from setting and we were getting to the end of our fuel. I was momentarily reveling in the exquisite sunset. There were small clusters of pink and gold cumulus clouds dotting the western horizon as the Vietnamese land mass slowly turned to a deep purple. I was so engrossed that I almost missed seeing the two junks!

They were big ones . . . the motorized kind . . . and I estimated them to be about 60 feet in length. Their sails were reefed and, judging from the size of their wakes, they were cruising at about 12 knots. They were about five miles inland and headed upstream in single file. Sailing into the beautiful sunset like that, they were a part of a gorgeous setting. It was sad, I thought, because in about 90 seconds neither of those pretty boats nor their cargo would exist.

We were out of time and fuel. This would be our first and last target. So, I decided, we would expend all of our ordnance on them. It was overkill, I knew, but it also made good sense.

"Firefighter Two, this is One. Two targets nine o'clock. Got 'em? Over."

"This is Two. Got 'em." was the rapid response.

"We'll make one run. Simultaneous," I directed. "I'll take the target farthest inland. You take the other. Runs will be parallel, north to south. Salvo everything and drive in close. Over."

"Wilco. Out." That was all the response I wanted as I added full military power and turned inland to the north of the river. I could make out a large tarpaulin covering cargo on the forward part of the deck and a small cluster of people gathered on the stern. I knew they were watching our approach and already regretting having started their logistics run south a little too early in the evening. Had they remained under cover just outside the mouth of the river half an hour longer, we probably would never have found them. Oh well, I thought. As an old academy professor used to say. "You pays your money and you takes your chances."

I felt like a cat that has just cornered a mouse. There was no place for them to hide. We approached a roll-in point just north of the river, still at 4,000 feet, with the speed building through 450 knots. George was now in trail behind me and would roll in on his target exactly when I rolled in on mine. I called him with a reminder.

"One rolling in. Switches on. We'll call each other's hits." I heard two microphone clicks in assent. What I meant was that I would watch his hits as I pulled out, and he would do the same for me. When shooting rockets, you can't watch your own hits . . . not if you want to survive.

It was a hard, nose-down left turn with about 120 degrees bank angle. I felt the comfortable squeeze of the g suit as I turned the corner. The airplane unloaded abruptly as I rolled out. It was a picture-perfect rocket run. There were about half a dozen black-clad people clustered in the stern of my junk. Each appeared to be firing a light automatic weapon. There were plenty of small yellow flashes. Good, I thought. This is the way I like it! Guns blazing!

The Crusader settled into a 20-degree glide angle and the speed began to build past 500 knots. I made a small, last-minute correction for lead angle on the gunsight and let the sight picture fill the windscreen. I was a little too low and too close

*Author lands A7A on USS SARATOGA (CV-60), Mediterranean, August 1971. (Official U.S. Navy photograph)*

On arrival at Cecil, Dick would get in my car, drive to my room in the BOQ at Cecil, help himself to a drink, change clothes and drive to his home in Orange Park, where he would spend the weekend. On Sunday evening, Dick would start the whole process in reverse whereby he would end up in the BOQ at Andrews late Sunday night and I in the BOQ at Cecil. The squadron preferred this arrangement to my keeping the airplane at Andrews because they always got their airplane back Friday night and had it all day Saturday. In all those months of flying an A-7 between Andrews and Cecil, we never got an airplane stuck there with maintenance problems.

Needless to say, both of us had the airways route structure memorized between those two places. The throttle was set at maximum continuous exhaust gas temperature and, I'm sure, the air route traffic controllers got to know our voices.

One Friday evening I was cruising up the coast on the airways looking forward to a pleasant evening with my family. About the time I was passing over the airways navigation Tacan station at Myrtle Beach Air Force Base, I spotted a gaggle of aircraft anti-collision lights far below me. I counted a dozen lights that seemed to be headed south and at a very low altitude. It was a clear, moonless night with plenty of starlight.

Suddenly I remembered that Air Anti-submarine Squadron TWENTY-EIGHT, a new member of my air wing, was moving from their homeport at Naval Air

Station Quonset Point, Rhode Island, to Cecil Field in a permanent change of station move. As a part of the CV concept evaluation, VS-28 was the first squadron of its type to join an attack carrier air wing.

Remembering also that I had signed the Operations Order directing the move, I became fairly certain that the bunch of transiting airplanes 20,000 feet below me was, in fact, VS-28. It was about the right time, I recalled, for them to be flying south.

I called the air route traffic controller on the radio and asked permission to make two left hand, 360-degree circles in place and to leave the frequency for communications purposes. I promised to monitor the guard (emergency) frequency. Traffic was light and they granted my request. I started flipping through my kneeboard, searching to find the Air Wing THREE radio frequency allocation card. I found it and, looking up VS-28's squadron common frequency, I dialed it in and listened. What I heard appalled me!

It was a steady stream of unprofessional radio transmissions indicating to me that this was probably the first time in the squadron's history that it had taken all 12 airplanes on a night, formation cross-country flight.

"Six, this is One, I don't have you in sight. What is your position. Over?" I recognized the Skipper, Commander Gerry Paulsen's voice sounding a little irritated.

"I'm at your four o'clock, Skipper. Watch out for that radio tower. It's 800 feet high," a voice responded.

"I see it, Six," Gerry's voice sounded a little frustrated. "Move it in a little closer."

"One, this is Five. I'm at your nine o'clock. I'm crossing my division over to your starboard side. Over," another voice announced.

Similar needless transmissions filled the night sky. I was unimpressed. But, I knew that Gerry Paulsen wanted to make a good impression, since he knew that most attack carrier community members were not exactly throwing out the welcome mat to the ASW intruders from the north. He would be appalled if he knew I were listening in. I couldn't resist the temptation to yank his chain a little.

"Dusty Dog, this is Battleaxe. Over," I transmitted. There was a rather pregnant, 10-second silence. All squadron transmissions ceased. Then Gerry's voice came on the air . . . a mixture of caution and incredulity!

"Battleaxe, this is Dusty Dog. Go ahead. Over." I could just imagine the look on his face, as he wondered where in the hell I was and how long had I been listening on this frequency to this sorry excuse for a night formation flight.

"Dusty Dog, this is Battleaxe," I continued. "Knock off all unnecessary radio transmissions and get some discipline into that formation. Over."

"Battleaxe, this is Dusty Dog. Wilco. Out," was Gerry's embarrassed response.

and recover four S-2s and two SH-3s every four and a half hours. At the slower pace, people can grab some sleep and thereby operate in that mode for long periods.

But Air Wing THREE still had two fighter squadrons, an EA-6A detachment and four A-6 tankers to operate. So we ran mixed operations with the ASW aircraft operating on the longer cycle and the jets on the shorter one, but at a greatly reduced tempo. It was the longest 10 days of my life. I recall finding the air boss sound asleep in his chair in primary flight control during launch and recovery operations. With all that din and roar, the poor man stayed sound asleep while his assistant ran the whole operation.

Of course, all of this was duly reported on in the daily traffic to higher authority, and great credit was claimed for what was described as a major breakthrough in carrier operations. I'm not so sure it was all that earth-shaking. But, it was a good exercise and did demonstrate, for the first time, that a CV could coordinate with the shore-based maritime patrol P-3s operating from Bermuda with a direct support submarine in effectively prosecuting simultaneous anti-submarine warfare and anti-air warfare operations in a multi-submarine threat environment for sustained periods.

There is one afterthought, a humorous one, that sticks in my mind, and gave me cause to wonder, in retrospect, at the fun quotient, or excitement factor, of sea-based anti-submarine warfare operations.

We were several hundred miles east of Bermuda when I launched in the pilot's seat of one of our S-2 airplanes on the eve of the great ASW cook-out. We flew to Bermuda for a big pre-exercise conference with the maritime patrol people. About an hour into the flight, the crew in back spotted something in the water and excitedly asked the squadron commanding officer (who was flying in the right seat) for permission to put a sonobuoy in the water. They wanted to put it near an object which I could not visually identify. Their Skipper looked across at me with an apologetic smile on his face and an inquiring expression.

He knew what it was and I did not. I nodded my agreement, more out of curiosity than anything else. Shortly after we put the sonobuoy in the water he explained that it was whale mating season and the crew in back had spotted a pair of whales "in flagrante delicto." We were about to watch a big bull sperm whale screw a docile cow. Not only that, I was told that if I threw a certain toggle switch in the large bank of them directly over my head, I could actually hear the sounds being made by the ardent swain and his lover as transmitted by the sonobuoy to our airplane!

After our airplane made two complete circles over the copulating mammals, it became obvious to the other three occupants of the airplane that I was not terribly impressed by whale screwing as much of a spectator activity. We pressed on to Bermuda.

Many of the regular air wing air crews were openly hostile to the intrusion of those "ASW pukes" into the hallowed bastion of attack carrier operations. I bent over backward to make it work, because it was the only air wing I would ever command, so it was going to be done well. But watching large sea mammals mate seemed a strange, but perhaps prophetic way to end the first evaluation of the CV concept.

# PART VIII

# THE END OF THE LINE

*"The quality of a person's life is in direct proportion to their commitment to excellence, regardless of their chosen field of endeavor."*

*– Vince Lombardi*

*This part is a sort of "last hurrah." The stories contained herein follow neither rhyme nor reason. Their only common thread is that they happened late in my naval career when meaningful things had ceased to be a regular part of my daily life.*

*Call it "aviator's menopause" if you wish. It doesn't really matter. Some retired senior naval officers will claim that the excitement continues right on up to the bitter end. I don't think so! A few of those who tried have even made the senior part of their service important. Some smaller number have even left their lasting mark in some way. But it isn't the same!*

*Maybe it only happens to aviators. Who knows?*

*But, until you startle yourself with a beautiful sunrise and try to preserve it by wrapping it in a little cocoon of intense emotional experience, or until you try very hard and achieve something very few people have ever done! Then you retreat to some inner recess of your consciousness and say to yourself . . . very honestly . . . I am pretty God-damned good!*

*Or, until you scare the living hell out of yourself and afterward conclude that, but for sheer luck, the air boss would be up on the flight deck saying to his flight deck crew as they look at the wreckage, "Let's clear this crap out of here. We've got some airplanes to land!"*

*Or, until you've sat in your machine . . . all alone, just you, at 40,000 feet in the inky black of your cockpit and said to yourself, "I can do things that most other human beings wouldn't dream of trying. I can do it with ease . . . with panache . . . with gusto and I can do it over and over again . . . and it's a piece 'o cake!"*

*Unless these sorts of things happen to you several times a week . . . month in and month out . . . you are no longer living the life to which your ego has become accustomed. When you stop living that kind of life, your existence has been forever altered and moved to a lesser level of abstraction. Try as you might, lie as you may! It just ain't the same!*

*In my particular case, the let down from an active flying career was gentled a bit by the numerous opportunities which came along to fly different flying machines simply because of the last few jobs I was fortunate to hold. Lots of people wanted me to fly their machines. It was amazing!*

*For example, the Northrop Corporation wanted me to fly their YF-17, the precursor to the F/A-18 Hornet. I did! Then those same people wanted me to fly their F-20 Tigershark. How could I say no? The people at McDonnell-Douglas offered me two flights in the Hornet. I did that too! Corky Lenox at Lemoore offered me two flights in the Hornet against the U.S. Air Force F-15. That was too much fun! My wonderful friend, Major General Phil Connelly at Edwards Air Force Base, arranged for me to go up there and fly both the F-15 and F-16. Before I was finished I had flown the F-16B, F-16 (J-79) and the F-16 XL as well as the F-15B. On one of those trips to Edwards, NASA gave me a ride in an old Lockheed F-104. Then there were the F-105 Thunderchief, the Bell XV-15 (precursor to the V-22) and even the twin-engine SH-2G and the British Hawk trainer.*

*But all of this notwithstanding, it was obvious my flying career was coming to an end. Different people handle the inevitable transition in different ways. Some do it well, others do it very poorly . . . and suffer for the failure.*

*As the Commodore said:*

*". . . I dare not ask for an improved memory but for a growing humility and a lessening cocksureness when my memory seems to clash with others. Teach me the glorious lesson that I may be mistaken . . . Give me the ability to see good things in unexpected places and talents in unexpected people. And give me, Lord, the grace to tell them so."*

# 36

## WALDO PEPPER

The most unforgettable scene in the movie "Waldo Pepper" was, in my opinion, the wing-walking sequence in which Robert Redford first climbed up onto the upper wing of a biplane and stood up. The camera captured the picture from his eyes. It was an eerie scene!

There was an ethereal sort of detachment as though the viewer were totally disengaged from anything tangible. All there was to see were billowy clouds, an endless, burning blue sky and, far below, the rolling green countryside of distant planet Earth!

The wing upon which he was standing was not in the camera's field of view. The upper arc of the whirling propeller blade was invisible, but the viewer knew it was there! One felt god-like, the master of all that could be seen of the realm of earth. Yes, the scene was truly unforgettable! I had, ever since, wondered what it would feel like to emulate Waldo Pepper . . . just once!

The 1974 annual open house and air show at Cecil Field was, by any standard, the most memorable air show I have ever attended. As the base commander, I was responsible for its planning and execution. A visit to an air base over a weekend by 120,000 visitors is no trivial event. It requires careful planning and the marshalling of substantial assets to ensure its orderly and safe execution.

Those assets, be they people, equipment or consumables, must be moved, monitored, operated and dispensed in strict adherence to well thought out plans, or chaos was certain to follow.

Fortunately for all of us, the event went well, the weather was perfect; and, much good will in northern Florida for the U.S. Navy as a result. But, it had its "down" moments!

*Author as base commander at Cecil Field, 1975.*

The grand opening of the show on Saturday morning began at 11:00 with a flair worthy of P.T. Barnum! I rode in the rear seat of a old World War II Grumman Avenger torpedo plane. The Avenger led a formation of vintage Confederate Air Force airplanes in a fly-by low over the reviewing stand. Sitting in the front row of that stand was my boss, the two-star Admiral from Naval Station Jacksonville, 10 miles east of Cecil.

He was, among other things, the most uninspiring naval officer I have ever known. But, like it or not, he was there . . . and, resplendent in his white, short-sleeved uniform, he was my boss! The fly-by was intended to impress him and to kick off the show with a bang. After the fly-by airplanes landed, the Avenger, my Avenger, chattering, sputtering, spewing black smoke, oil and hydraulic fluid, stopped and shut down its engine directly in front of the reviewing stand. I vaulted out of the rear cockpit, colorful in my "Sierra Hotel" flight suit and garrison cap at the correct jaunty angle, ran up to the reviewing stand, grabbed the microphone, introduced myself and welcomed everyone to the annual air show!

"Good morning, ladies and Gentlemen, I'm Captain Gillcrist, Cecil Field's base commander. On behalf of my boss, Commander Sea-based Anti-submarine Warfare Wings, Atlantic, I welcome you to our annual open house and air show. Let the show begin!"

There was much clapping and cheering from the spirited audience as I walked over to the Admiral, saluted and took the seat reserved for me at his immediate left. There was a sour expression on his face. He hadn't returned my salute, so I sat down beside him. "Well, Admiral, how did you like that?" I asked brightly. He responded dourly.

"It's not a very mature way for a senior naval officer and one of my base commanders to behave." I felt as though a bucket of ice water had been doused over my head. Then he seemed to turn his attention away from me toward the first airshow event. To hell with this, I thought and, getting up, I walked down the line of parked dynamic display aircraft. My two sons followed me. I must have looked pretty dejected.

"Hi, Skipper. How's it going?" I looked up and recognized an old friend, Joe Hughes, the pilot of our wing-walking act. He was to perform later. Joe was a sight to behold! He stood there next to his Super Stearman biplane, a roguish expression on his face, white-haired and sporting a huge handlebar moustache.

"Okay, Joe," I answered. "I just wish my boss had a better sense of humor." That, I thought, was putting it mildly! He had no sense of humor whatsoever!

"Don't let it get you down," he said, adding with a twinkle in his eye. "Come on. Get a pair of sneakers on and I'll take you for a ride!" He gestured with his head toward the upper wing of his Stearman where the wing walking rig was, and I got his drift right away.

"How, when?" I asked.

"As soon as you can get back here with your sneakers, and as soon as I can get the FAA official to fit us in between events in the schedule. And right on top of the wing where you'll get a great view!"

Without a second thought, I made a decision. "You're on, Joe. I'll be right back." I headed for my official car at a trot. I kept my gym gear in the trunk. In my gym bag was a pair of sneakers. My two sons, 11 and 14 years old, stood there, mouths agape, eyes as big as saucers, wondering what their father had in mind. In less than two minutes I was back, wearing the sneakers and admonishing the two youngsters, "Not a word of this to your mother!"

None of the crowd seemed to notice me standing on top of the Stearman's upper wing. Joe strapped me to his airplane with foot straps and another around my waist and to the bar, guyed up on the upper wing by wires. The crowd was occupied with watching a stunt plane do "lumpshevaks" with white smoke streaming from the wing tips. They paid no attention to us. Joe handed me a black cloth helmet and goggles which I put on, and we were ready to go. He told me the FAA man had approved a takeoff right after the stunt plane presently performing. There wasn't a moment to lose.

As Joe started the engine, I reviewed what I was about to do, for better or worse. "What am I doing?" I asked myself. Here I am, a 46-year-old senior naval officer, the base commander, and I am about to make an ass of myself in front of my boss, my two sons and, by the way, 60,000 spectators. Oh well, I decided. If the Admiral thought I was immature five minutes ago, wait 'til he sees this! I'll show him just how immature I can be when I try!

My second thoughts were interrupted by the cough and then roar of the Stearman's engine . . . too late now! Joe's only instructions were to keep my hands by my sides during the landing.

"They really screw up my flight controls at low speed," he explained. We were now taxiing slowly down the taxiway to the southern end of the north-south runway. The two tiny figures of my sons, still watching me aghast, were diminishing in the shimmering heat waves coming off the hot concrete taxiway.

We idled a few moments in the warm-up area at the end of the runway. I looked down and behind me, feeling terribly stupid and foolish, and noticed Joe talking on the radio through his lip microphone. He looked up at me and, as he did so, flashed that broad, insouciant smile and flipped me the well-known thumbs up signal. Then the engine roared and we taxied awkwardly out onto the runway. Here I am, I thought, about to do the stupidest thing in my naval career! The engine came to a shrieking roar, the tips of the propeller blade just a foot or so in front of my knees, and up ahead, 60,000 spectators, my two sons and a dour-faced Admiral all watching me! "I think I have lost my frigging mind," was my last thought as the wind whipped my face harder and harder. We were rolling.

I felt the tailwheel lift off the runway. Moments later I rose from Mother Earth as the reviewing stand flashed by my left side and beneath me. Breathing was difficult. Joe's only other word of advice had been to keep my mouth shut. What I had thought of as a joking remark was actually, I realized, very serious advice. A 120 miles-per-hour gale was now blasting at me, tearing at my goggles, helmet and flight suit and trying to literally rip them off. A random thought crossed my mind. Joe had advised me what not to do. But had he ever ridden up here himself? I was now certain the answer was no. He couldn't be that stupid, I concluded!

But another sensation now had the upper hand. In fact, it was so powerful, so compelling, that all other sensations paled to insignificance!

I felt god-like, immortal, ethereal! Joe had leveled off and throttled back at about a thousand feet. I was guessing, because with this new perspective it was difficult to judge. We were cruising at probably 90 knots over the green forested area to the southwest of the field. The scattered clumps of fluffy white clouds were much fluffier and whiter than I had ever seen them before from the safety of a cockpit. The sky was unbelievably blue . . . I wondered why? I don't recall feeling the wind blast at all!

Looking straight ahead, there was nothing in my field of regard but the earth and sky. When I die and if I go to heaven, I thought, this must be what it feels like. The beauty of the scene, and the sense of floating all alone in the sky was mind-boggling! I recall thinking, "Oh God, if only this could go on forever!"

*Author wing-walking at Cecil Field Air Show, 1975. (Author's collection)*

The plane banked sharply. I looked down and back at Joe. He returned my gaze and pointed his hand down and back toward the field.

I correctly guessed that he had spoken to the tower and we were being called back for a landing during another break in the air show events. I remember being disappointed that it was all going to end . . . much too soon! Joe throttled back and we began a gentle descent back toward the field to what I assumed would be a landing. But something was not right! We were much too high for a landing. We crossed over the end of the runway a good seven or eight hundred feet over the ground.

I recall being perplexed but not overly concerned until I saw the twin streams of red and blue water vapor coming from the upper wing tips! From the left tip it was bright red. From the right it was blue. It's funny, the things one remembers at moments like this. I knew immediately that I was in trouble. Joe never mentioned anything about this! He obviously had something in mind that he hadn't bothered to tell me about.

The nose of the Stearman dropped abruptly. For a moment my stomach floated up to a point somewhere between my lungs and I almost donated its contents to Cecil Field's ecology. The airplane was now in a dive angle of about 45-degrees and pointed at a spot on the runway directly in front of the reviewing stand, in front of 60,000 spectators, my two sons, and my boss!

Of course, from my perspective, the dive angle looked more like 90 degrees! The speed increased to what I guess must have been 175 or 180 knots. I couldn't breathe. My mouth was being torn open. My lips were splayed wide open despite gritted teeth. My nose felt like it was being jammed into my head. The only sense I had that was undistorted was my vision.

Just before we were about to plow into the concrete, Joe hauled back on the stick and rammed full power to the Stearman's engine. The airplane seemed to bottom out 10 feet off the runway and started up into what was an almost vertical climb.

As we bottomed out a brilliant idea came to me. I had already braced my legs for the two or three g pull out and decided to send my grouchy boss a signal. Turning my gaze to the reviewing stand, I spotted the Admiral sitting there just a few hundred feet way. Then, just as the airplane bottomed out of its dive, I flung my left arm out to full extension and flipped him the bird!

The wind blast nearly yanked my arm from its socket, but I held the international signal for the requisite two or three seconds . . . there must be no mistaking it, I decided.

God, I thought! What a stroke of genius! What a tremendous feeling that simple gesture, made so famous by an ebullient Nelson Rockefeller, made me feel! My day had been made!

Joe made a right turn to the downwind leg of a normal landing at the southern end of the field. As we made the long taxi back to where he was parked, I looked down and back at him. Joe was laughing so hard that tears were streaming down his face and into that magnificent moustache. It was a contagious laughter. I joined him. By the time I had unstrapped and climbed down from the wing we were both in an uncontrolled state. My two sons stood there gaping again at their father. Clearly he had taken leave of his senses. My two sons accompanied me back to my car where I swapped my sneakers for flight boots and admonished them a second time not to tell their mother. Then I returned to the reviewing stand where I expected to be told of my impending courtmartial for having made an obscene gesture at the Admiral before 60,000 witnesses. He gave me a brief glance as I took my seat next to him. In that glance I knew! In that split second when our eyes met I knew, beyond any doubt, that he never recognized the helmeted and goggled mad man in the black flight suit who had just flipped the bird at the crowd, at his very own base commander! What a wonderful feeling!

14 to McDonnell-Douglas in St. Louis for a two flight orientation. The orientation, tailored for visiting firemen from the Pentagon, included a two-hour simulator, a one-hour demonstration in the back seat of a two seat F/A-18B, and a one hour flight in the front.

The first flight landed at nearby Whiteman Air Force Base, where we gassed up, switched seats, had a quick lunch and took off. The final landing was back at St. Louis. The two flights were intended to give the newcomer a proper appreciation for the weapons system, strike and fighter capabilities and the flying qualities and performance of this remarkable airplane. For me the orientation accomplished all that and more.

The Hornet had one drawback which was well understood from the outset of the program. It didn't carry enough internal fuel. But the drawback was not as serious as most people thought. The decision not to increase internal fuel capacity was made to ensure that the controversial airplane not fall short of two very important specifications; approach speed and acceleration. The addition of internal fuel would most certainly have caused the airplane to broach one or both of those specifications. In retrospect, it was probably a good decision.

The internal fuel capacity proved to be a much more serious political problem than an operational deficiency. The airplane's fully automated leading and trailing edge flap system made it an entirely different airplane when compared to conventional designs. Consequently, initial evaluations by operational testers were not favorable. However, the fleet operators were, at the same time, amassing many operational flight hours in the machine and a great deal more expertise than the operational testing community. They were learning how to utilize its designed performance characteristics and were achieving substantially greater combat radii with the airplane in realistic operational scenarios.

The Hornet had enemies within the Navy. A large and vocal part of the medium attack community, for example, feared that the arrival of the Hornet would relegate them to a far greater aerial refueling role than they preferred. This paranoia resulted in a serious attempt to kill the Hornet from within . . . and it nearly succeeded.

The first operational evaluation of the F/A-18 was orchestrated by a member of the medium attack community and articulated by a surface warfare flag officer who was woefully unaware of the "hidden agenda" and dismally ignorant of Naval aviation requirements. The result was a preordained disaster. Congressional tinkering only served to compound the problem.

The perceived "schism" in the blue-suit Navy was exploited by the press and several other groups with axes to grind. Fortunately, the fleet operators, Marines and Navy, came to the rescue of the beleaguered few in Washington who were desperately trying to serve the fleet's needs. The Hornet ultimately made it through

*Author in cockpit of Northrop YF-17 at Edwards AFB, January 1980. (Author's collection)*

*Author checking out in F/A-18B NAS Lemoore, California, 1979. (Official U.S. Navy photo)*

the acquisition quagmire and proceeded to set records for fleet introduction un-matched by any other airplane in history.

The true operational utility of the Hornet can best be illustrated by an event which occurred to me which brought home to me, in the only terms which are meaningful, its true value to the fleet.

The telephone in my office in "Fightertown" rang one morning in May 1981. The voice on the other end was that of my old friend, "Corky" Lenox, my func-tional wing counterpart at NAS Lemoore, California. He was responsible for the care and feeding of the Pacific fleet's strike fighter community. He had called to invite me to come to Lemoore to fly his F/A-18. It was truly "his" F/A-18 because, for several years, he had been the airplane's program manager in Washington. I accepted his offer without giving it a second thought.

It was only three weeks later, almost to the day, when I landed my Tomcat at Lemoore . . . ready to fly. Corky was waiting at the foot of the ladder, hand out-stretched and a broad grin on his face.

"Welcome to Strikefightertown," he said as we proceeded to his waiting se-dan. "You are going to like what we have prepared for you today," he said with a hint of a smile lingering at the corners of his mouth.

The first of two flights that day was in a two-seat F/A-18B with the C.O. of the fleet replacement squadron, Commander Jim Partington, riding shotgun. Jim proceeded to demonstrate the airplane in its air-to-ground mission. Two things impressed me immediately; the weapons system and the airplane's agility in the strike environment. The net effect of being able to bomb with unprecedented accu-racy on the first pass and to achieve, through agility alone, minimum exposure to the surface-to-air threat represents an enormous increase in survivability. That les-son came through loud and clear on the flight with Jim Partington, and I was still digesting it when we walked into the briefing room for the second flight.

The reason for "Corky's" mysterious smile became clear to me when I opened the briefing room door and saw who was inside waiting for me. There were two young men in Air Force flight suits sitting there smiling. We were introduced, and the flight leader informed me that they were from Holloman Air Force Base. The mission, I learned, would be two-versus-two air combat maneuvering against the two F-15 Eagle pilots. This, I decided, was going to be fun. A young fighter pilot named Scott Ronnie would ride shotgun with me and the other Hornet would be a single seater.

The four airplanes, each configured with a single centerline external fuel tank, took off and headed east over the mountains to the Mojave Desert restricted area where unrestrained air combat maneuvering was conducted.

Once in the center of the assigned area our two sections of airplanes split up. The Eagle pilots headed for the southeast corner of the range while out two Hor-

*Flight of F/A-18s over the Sierras, NAS Lemoore, California, 1984. (Official U.S. Navy photo)*

nets went to the northwest end. On Scott's call the two sections turned back to-wards each other. About 40 miles separated us when we came out of our turns. Our two Hornet pilots agreed not to try any fancy section tactics, but to meet head-on with our opponents and then provide mutual support.

About a minute out of our turn, we picked up radar contact co-altitude with us at 15,000 feet. Closure rate was almost a thousand knots. Since we were poking along at what the Hornet pilots called their best "cornering velocity" of 340 knots, I knew the F-15s had to be doing over 600 knots. This cornering velocity stuff ran counter to everything I knew about aerial combat and energy maneuvering . . . where speed is still life! However, I agreed to try it. I'll try anything . . . once!

I made visual contact at 15 miles because of unusual lighting and background contrast. The F-15s were leaving black smoke trails which I could pick out easily. I felt absolutely obscene staying at 340 knots with the enemy, now in full view, doing almost twice our velocity.

The two sections passed down each other's right sides. My airplane, on the right side of our section, passed just a hundred feet or so from the nearest Eagle. Just before he passed abeam, he "broke" into me in a screaming climbing turn. I ran the throttles to full afterburner and matched his move. Someone shouted, "The fight's on," over the radio, and indeed it was!

After about 90 degrees of turn, the Eagle reappeared into my view over my right shoulder and about 5,000 feet above us. That didn't surprise me, given their higher energy level. What did surprise me though, was the fact that I had already gained about 15 degrees advantage on him already. By the time we had completed a full 360 degrees of turn, he was still about 5,000 feet above me but I now had a good 90 degrees advantage on him.

I'm sure he realized the trouble he was in, because he tried to pitch down into me . . . and meet head-on. Surprisingly, I was still at cornering velocity and came out of afterburner while, at the same time, reefing the Hornet into heavy buffet. The F-15 came by me headed straight down and passed me, causing a little over-shoot . . . but not too much. He tried to take advantage of my overshoot by revers-ing his turn and starting a scissors engagement. It didn't work and I slid in behind him.

I called, "Guns, guns, guns!" My quarry came out of burner, rocked his wings and called. "Knock if off!"

That was the signal to stop the engagement. There were two more engage-ments, each one almost a carbon copy of the first. At the "pass," each time the flight degenerated into two separate one-versus-one engagements. I didn't see much of the other two airplanes except for some radio transmissions which were accom-panied by some heavy grunting. Our opponents used afterburner much more than we did. As a consequence, after the third engagement, as someone called "Knock it off," my adversary called that he was at "bingo" fuel state. The two F-15s both headed back for Lemoore.

I was totally astonished! Here I was, flying an airplane that had been highly criticized (almost canceled) for its lack of internal fuel capacity (we use the term short-legged). And I just ran the highly touted, long-range F-15 out of fuel! Amaz-ing! The real operational significance of the Hornet was beginning to sink in!

Our section of Hornets each had enough fuel remaining to have one good go at each other after the Eagles had departed for Lemoore. I think the other Hornet pilot let me win the last fight. He was being very nice. Thirty minutes later we taxied into the line at Lemoore and I had that good, tired, soaking-wet feeling of having done the one thing I'd rather do than anything else in the world.

The Hornet is one hell of an airplane! Someday some base commander is going to mount a Hornet on a pedestal at his main gate. The inscription on the plaque had better say "The Main Battery of the Fleet!"

But there needs to be one very serious caveat added to what has been stated above. The Hornet is a wonderful airplane as long as it is employed for what it is . . . a wonderful strike fighter representing the low side of Naval Aviation's "high-low" mix!

It turns out, unfortunately, that in 1990, the Navy embarked on an extremely expensive program to make a silk purse out of a sow's ear! By 1995 they will have spent over ten billion dollars trying to make the Hornet something it could never be . . . an advanced strike fighter!

At a time in history when Naval Aviation should have been developing the new, stealthy, multi-role, deep-interdiction, advance strike fighter, its leadership simply "lost the bubble." Instead, they wasted the taxpayer's money with a degree of profligacy beyond belief. Naval Aviation may never recover from this disastrous mistake, because it may ultimately result in the end of carrier aviation as we know it today!

# 38

## TIGERSHARK

O f all of the tactical airplanes I have ever flown, the best by far for close in combat aerial maneuvering was the ill-fated F-20 Tigershark. The F-20 was also one of the most controversial aircraft development programs ever begun. An independent, private-development effort, the Northrop F-5G (later re-named the F-20) was the corporation's response to President Carter's FX program.

The FX was intended to be a low-cost tactical airplane for the export market, but fully supported by the U.S. government and the U.S. Air Force. Therein lies the root cause for the failure of the concept and the ultimate termination of the program. The F-20 and the FX actually got support from neither the U.S. Government nor the Air Force. The Air Force did not want the F-20, and until they bought it, no foreign customer would dare buy it for fear of lack of follow-on support.

The U.S. Air Force had not wanted the F-20's predecessor, the F-5, but it had been forced upon them by influences within the federal government. Once in their inventory, however, the F-5 served the Air Force extremely well in the adversary role.

Also, since the Air Force had bought it, albeit in limited quantities, the F-5 became a very popular airplane in the export market. Some 2,500 of them were sold overseas. At the time the FX program was kicked off by the Carter administration, there was a definite need to replace the aging F-5s with a newer, more capable airplane.

The FX was going to fill this requirement, and the Northrop Corporation proposed the F-20. In competition, the General Dynamics Corporation proposed a stripped-down version of the Air Force's popular front line fighter, the F-16 Fal-

*Author's second flight in the Northrop F-20 Tigershark at Edwards AFB, 31 July 1984. (Author's collection)*

con, developed with Air Force research and development dollars and produced in an Air Force-owned factory. Northrop, on the other hand, spent about $1.2 billion of company money before finally acknowledging that they had violated a basic rule of the defense industry acquisition game: "Never try to force a program down a customer's throat!"

Two factors helped lead the corporation down this particular "primrose path." The first was their success in forcing the F-5 upon a reluctant U.S. Air Force. This factor became a two-edged sword. First, it established within senior members of the Air Force a "never again" mind set against future Northrop "capers." It also encouraged decision makers within the corporation with a false sense of encouragement ("We did it with the F-5, so we can do it with the F-20 as well!").

The second factor which helped lead to Northrop's failure was promised "help" from the White House which they never obtained. As a result, the U.S. Air Force, aided and abetted by General Dynamics, defeated Northrop at every juncture in the FX program in the acquisition path.

As a sub-plot of the main drama, the U.S. Navy was quietly building support in the Defense Department and on Capitol Hill for a modest adversary aircraft acquisition program. A line item had been in the President's budget for several years for 24 "adversary airplanes" in the mid-1980s. Unfortunately, the dollars were never obligated because the F-20 versus F-16 controversy was never resolved. For the Navy, the acquisition program required an off-the-shelf procurement. As long as the FX controversy percolated along, the issue would never be resolved. Key Navy officials went to the extent of flying both contestant airplanes to make a technical judgement.

When the U.S. Navy finally established a bone fide adversary requirement of 48 airplanes in 1985, a competition was initiated and, because of what Northrop described as "predatory pricing," General Dynamics "gave away" the Navy's F-16N adversary airplanes for a song. The Navy was pleased at the bargain they had struck . . . and General Dynamics was especially pleased at having finally driven a wooden stake into the heart of the F-20 program . . . once and for all!

Hindsight is always easy. But viewed in that perspective, it is easy to conclude that the Northrop Corporation made a monumental miscalculation, a $1.2 billion miscalculation for which the stockholders should have ridden the corporate leadership out of town on a rail, appropriately tarred and feathered!

The final chapter of the F-20 program was the Air Defense Fighter program. This was a program calling for the acquisition of enough aircraft to replace all of the "cats and dogs" which comprised the Air Defense Command's wide ranging inventory of fighter interceptors. This, a very large program, was essentially a competition to select the fighter which would replace the aging inventory of machines which made up the Air Defense Command with an interceptor "not as expensive as the F-15."

Again, the two finalists were the F-20 and the F-16. Although there was a modicum of support within the independent thinkers in the Air National Guard for the F-20, the "regular" Air Force and General Dynamics team were now unbeatable. Again, the customer "didn't want the product."

The final straw for both the FX concept and the F-20 program was the announcement by the Air Force on 31 October 1986 that the F-16 had won the Air Defense Fighter competition. That was the day the F-20 died!

But long before that occurred, when I was Director of Aviation Plans and Programs for the Navy, I had the distinct privilege of flying all of the competitor airplanes, including the Tigershark! My evaluation of the F-20 consisted of two flights on 31 July 1983.

To be on the safe side, Northrop required that anyone who flew the F-20 had to have completed "a few" flights in the F-5 in the previous few days. The aircraft characteristics and cockpit geometry of the F-5 were similar enough to constitute a good "warm up" airplane.

I flew to Edwards Air Force Base after a few warm-up flights in a Topgun F-5, and was given a "check ride" in a Northrop F-5F by Northrop test pilot Dave Barnes. Dave politely called it a warm-up flight. In fact, it was a check flight to ensure that it would be safe enough for the corporation to entrust their several hundred million dollar investment into my hands. Dave apparently gave me an "up!"

That same afternoon, Daryll Cornell gave me a two-hour briefing on how to fly the Tigershark. He was a quiet, soft-spoken man for whom I developed a great deal of respect. He was the one who would "chase" me for my two flights.

The next day, bright and early, we conducted a pre-flight briefing on the course rules for the Edwards restricted area and the sequence of events for the two-flight program.

The first flight would be mainly orientation and a look at the maneuvering envelope of the airplane. The second flight would be a full-afterburner scramble takeoff to altitude with a supersonic dash to the limit of the flight envelope thus far achieved in the on-going flight test program.

Daryll and I approached the airplane for a pre-flight inspection and walk around. It was a beautiful looking little machine! It sported a jazzy red-and-white paint job and had the look of the future in its lines. It also had "air superiority" written all over it!

The Tigershark had a wingspan of just 35 feet and was a little under 60 feet long. It had an empty weight of only 12,400. With a full bag of 5,200 pounds of internal fuel, it weighed just over 18,000 pounds in the fighter configuration with a Sidewinder missile on each wing tip. If it carried two external fuel tanks it could achieve a ferry range of 1,700 miles.

*Author (center) on the occasion of his F-20 flights. Northrop's Chief Test Pilot Darryl Cornell stands at the right, and Chuck Johnston is at left. Photo taken on 31 July 1984. (Author's collection)*

One of the principle differences between the F-20 and its predecessor, the F-5, is that it had a single engine. It was powered by the highly successful General Electric F404 turbofan engine which had been tweaked up to produce 18,000 pounds of thrust in full afterburner. That engine could boost the little fighter out to twice the speed of sound. At a combat weight of 50 percent internal fuel the F-20 had a thrust-to-weight ratio of 1.14:1, which gave it tremendous acceleration and climb rates.

My first flight began with a takeoff in military power (no afterburner) and a climb to 20,000 feet where I did mild maneuvering to get the feel of the airplane. Next, I examined its stall characteristics in both the clean and landing configurations. I found that it had tremendous flying qualities and handling characteristics, and honest stall warning and post stall behavior. We then climbed to 30,000 feet, kicked up the speed to 500 knots and did some serious air combat maneuvering. The airplane's honest flying qualities in this regime engendered high pilot confidence almost immediately.

We started back to the field and did a series of acceleration runs in level flight at several altitudes. Each time the F-5F chase plane would pull up alongside me and we would stabilize at 200 knots. Then, on signal, we would both go to full power, or full afterburner and observe the airspeed indicator. Each 50 knots I would call out airspeed and Daryll would do the same.

The acceleration of the Tigershark was eye-watering and, at all altitudes my airplane literally ran off and left the F-5 in its dust.

The final part of the first flight was an examination of the F-20's handling characteristics in the landing pattern. I shot a series of carrier landing approaches and liked what I saw very much. The final landing of the first flight was followed by a working lunch at which we debriefed the flight with the engineers and briefed for the second flight.

I was really looking forward to the scramble flight because everything was going to happen in a blur of speed. After lunch we again manned up and this time we knew that Daryll would be left far behind. The first flight had only lasted 48 minutes because of the low altitude afterburner acceleration runs. The second flight, if it followed the script, would only last 36 minutes.

For the second flight the plan was to make a full afterburner takeoff, climb to 38,000 feet, accelerate out to 1.6 times the speed of sound, come out of afterburner at that speed and do some supersonic maneuvering at that altitude.

As I taxied onto the duty runway at Edwards, I found that my pulse was quickening. Clearance for a high-performance takeoff was received and I held the brakes and advanced the throttle. Time seemed to stand still for a moment as I looked down the runway, through the shimmering heat waves, across the valley floor and up the slopes of Alamo Mountain, all the way to its 7,000 foot crest. The air was so clear that I thought I could almost distinguish the trees on its slope (an easy 40 miles away). This dry lake bed which had seen so much aviation history, was about to be my launch pad for the wildest ride of my life!

The engine rpm had reached full military power. My pulse was really racing as I simultaneously advanced the throttle to full afterburner, released the brakes and punched the time clock on the instrument panel in front of me. What followed is recorded on a tape hooked up to the radio in my oxygen mask. I talked to it to record data points.

The afterburner light was a solid shove in the small of my back, and the acceleration down the runway was startling. Everything seemed to come at me quicker than I anticipated. The airplane reached nose wheel lift-off speed in a few seconds and literally jumped into the air. I held it level at about 50 feet off the runway and snatched up the wheels and flaps. Optimum climb schedule was 0.9 Mach and 300 knots was the speed to begin programming the stick back to rotate the airplane to a 45-degree climb angle on the attitude gyro. I did it a little too slowly and the Mach stabilized at 0.94. I tried to raise the nose just a little to get it back to 0.9 and it didn't seem to want to slow down. When the attitude gyro showed a climb angle

*Flight of F-20s over Nevada desert. (Author's collection)*

of 50 degrees I quit trying for fear of overshooting and really screwing up the climb profile.

I began a gentle turn to the right to stay within the restricted area boundary and noted what I call "blue vertigo." With nothing in the pilot's field of view but blue sky, he feels deprived of normal ground reference points. The altimeter needles were whirling wildly and I went through 20,000 feet feeling like I was sitting on the front end of a rocket.

I realized as I passed through 30,000 feet that I needed to anticipate the level off at 38,000 because it meant lowering my nose attitude much more than I was accustomed to. Even though I anticipated the level-off, I was a little slow and had to push the airplane through the last 10 degrees rather abruptly, causing a small zero g hump. The sweep second hand on the stop clock read two minutes and twenty seconds from brake release.

Seconds after level-off, the Tigershark went supersonic. After that, the Mach meter needle marched inexorably up the dial until precisely at 1.6 I came out of afterburner and started a gentle turn to the left.

Two things surprised me. If it had been a Phantom or even a Tomcat, I would have felt my body thrown forward against the shoulder restraints by the deceleration. Not so in the F-20! Secondly, as soon as the turn started in either of those other two airplanes, deceleration would have been greatly increased. Not so in this machine! Deceleration coming out of burner was almost gentle, and going into the turn was surprisingly slow.

I called starting my reversal turn to assist Daryll in joining up. He told me he had me in sight, but was 20 miles behind and 20,000 feet below me! I completed the 180-degree turn still supersonic and, pulling up the nose, did a couple of aileron rolls . . . still supersonic! Finally, I pulled the throttle to idle power to help Daryll in his rendezvous.

All of this meant, of course, that the Tigershark had very clean lines and what I call "supersonic persistence." As Daryll joined up, I noted my fuel quantity and was astonished to see that I had burned only 1,200 pounds of fuel to get all the way to 1.6 Mach at 38,000 feet. I guess that my fuel consumption to do the same thing in a Tomcat might be five times that much!

We did some high-altitude maneuvering turns and then began our descent to do the final demonstration high g maneuvering at low altitude and a high thrust-to-weight ratio.

As briefed, I leveled off at 2,000 feet above the valley floor and decelerated to 300 knots. I then ran the throttle to full afterburner and began a left climbing turn, steadying up at 500 knots and holding a constant five gs. The amazing airplane was holding these parameters steady and still climbing so fast that the rate of climb needle was pegged at 6,000 feet per minute. "Holy cow!" I thought. What a maneuver that would be in a dogfight! Right at that moment I calculated my thrust-to-weight ratio to be almost 1.2:1!

The final landing at Edwards was uneventful. As the airplane rolled out on the runway I began to realize that I had just flown one of the best-engineered tactical airplanes ever built! It had better flying qualities than any airplane I had ever flown. Taxiing back to the line, I decided that this airplane had better control harmony across the flight spectrum than any airplane I had ever flown. I recall thinking, "This one is a winner!"

As a footnote, there is an eight by ten-inch color photograph of me standing in front of that beautiful airplane. It is the kind of picture that pilots normally frame and hang on their "I love me" wall in their den. But that will never happen to this picture. Standing on either side of me are Daryll Cornell and Dave Barnes, both great guys! There are broad grins on all our faces.

The reason I'll never display the picture is because Daryll was killed doing an air show in Suwon, South Korea, on 10 October 1984. Dave Barnes was killed on his way to the Paris Air Show while practicing his routine. The crash occurred at Goose Bay, Labrador, on 14 May 1985. Almost as sad as those two untimely deaths is the fact that a wonderful machine, and an engineering marvel never went into production!

# EPILOGUE

*"In the end, the only people who fail are those who do not try."*

There comes the right time to end a career as a carrier pilot. Most of us don't realize that it was the right time until viewed with the unerring aid of hindsight. In my case, it was the moment I was fitted out with a pair of artificial hip joints made out of nothing less than high-grade titanium.

To be sure, I did some flying after that (including the Tigershark), but the realization was there at that time that I would never again land an airplane on an aircraft carrier. In fact, I instinctively knew it the day I made my last carrier landing in an F-14 on USS Kitty Hawk 21 October 1981.

While sitting in the cockpit preparatory for my last catapult shot and the return flight to NAS Miramar, I took in all the familiar sights, sounds and smells for one last time.

Somewhat overcome with the sense of the moment, I scribbled some thoughts on my pilot's kneeboard. Later that evening I committed them to a clean piece of paper. They make a better end for this volume than I could do with prose.

### ". . . At Forty Thousand Feet"

*"It matters not how wild the storm*
*Whose tow'ring seas and stifling gales*
*Intimidate the bravest few,*
*Who dare, upon the seas, to sail.*

*For deep inside that vaulted dome*
*Of Stygian clouds and roaring winds,*
*Of pelting rain and frozen sleet;*
*There plows a doughty battle fleet.*

*Despite the awesome show of strength*
*Which gods select those ships to test,*
*Their flight decks spew forth fragile craft*
*Those mighty elements to best.*

*And in those tiny planes' ascent*
*The long and tortuous climb begins.*
*The maelstrom pales from black to slate;*
*The mighty blows do slowly 'bate.*

*And, bursting forth as in a dream,*
*Into a world all bright and clean,*
*A truth reveals this small elite,*
*(The youngsters from the battle fleet);*
*The sun is always shining bright*
*On those at forty-thousand feet.*

*The blackest night of overcast*
*Keeps stars and planets, and the moon,*
*From helping pilots coax their craft*
*From catapult into the gloom.*

*Each tiny cockpit dimly lit*
*With dials and needles, the path to show.*
*The pilot claws for speed and height*
*Above the deadly sea below.*

*But startling as that dark domain,*
*Recedes up through the mist and rain,*
*Those dauntless craft emerge reborn*
*And from that ragged dome are torn.*

*A million stars light up their craft.*
*The moon and planets bathe in light,*
*Their cockpits glow in warm delight,*
*The awesome spectacle of night.*

*But, as in life, sojourn is brief;*
*And all those airmen need attest,*
*An hour or so will scarce complete,*
*Before those birds must come to rest.*

*Down, through the storm back to the womb,*
*Into the inky cloud they grope.*
*Their dials and needles the only hope*
*To thread the needle and arrest;*
*The price for the bird's return to the nest.*

*But it matters not how wild the storm,*
*How dread the seas of life around,*
*How dire the fate and close the doom*
*As doubts and fears so quick surround.*

*The few recall, with vision grand,*
*(The fliers from the battle fleet);*
*That stars are always near at hand*
*For those at forty thousand feet!"*

*Author checks out in Northrop F-5A from TOPGUN, August 1979. (Author's collection)*

*Author taxiing out in NASA's F-104 at Edwards AFB, October 1979. (Author's collection)*

*Author checks out in USAF F-15B at Edwards AFB, November 1979. (Author's collection)*

*Author in F-16XL (left) Edwards AFB, November 1979. (Author's collection)*

*Opposite below: Author (left), LtCol Dave Milam, MajorGen Conley, and protocol officer Adrian Jowyk at Edwards AFB, November 1979. (Author's collection)*

*Einar Enevoldsen, NASA Chief Test Pilot, and the author after check out in Lockheed F-104 Starfighter. (Author's collection)*

*Author checks out in F-16 (J-79) at Carswell AFB, June 1980. (Author's collection)*

*Author and LtCol Dave Milam discuss a fine point on the F-16 checklist at Edwards AFB, February 1980. (Author's collection)*

*Author checks out in Fresno ANG F-106, NAS Miramar August 1980. (Author's collection)*

*Author makes his first F-14 carrier landing at age 51 aboard USS KITTY HAWK (CV-63) on 21 October 1980. (Author's collection)*